Diesel Retrofits: Quantifying and Using Their Emission Benefits in SIPs and Conformity

Guidance for State and Local Air and Transportation Agencies

Transportation and Climate Division
Office of Transportation and Air Quality
U.S. Environmental Protection Agency

United States
Environmental Protection
Agency

EPA-420-B-14-007, February 2014
Supersedes EPA-420-B-06-005

SECTION 1: INTRODUCTION — 6
  1.1 What is the purpose of this guidance? — 6
  1.2 How can emission reductions from retrofit projects be used to meet SIP and conformity needs? — 7
    1.2.1 On-road and nonroad retrofit projects in SIPs — 7
    1.2.2 On-road retrofit projects in transportation conformity — 8
    1.2.3 Nonroad retrofit projects in transportation conformity — 8
    1.2.4 Retrofit projects in general conformity — 9
  1.3 What vehicles, engines, and equipment does this guidance address? — 9
  1.4 What is a retrofit project? — 9
  1.5 What are the requirements for using vehicle, engine, or equipment replacements in SIPs and conformity? — 10
  1.6 How does this guidance relate to the Voluntary Mobile Source Emission Reduction Program SIP guidance? — 11
  1.7 Can a state require retrofits of on-road and nonroad vehicles, engines, or equipment? — 12
  1.8 Does this guidance create any new requirements? — 13
  1.9 Who should I contact for additional information? — 13

SECTION 2: QUANTIFYING RETROFIT EMISSION REDUCTIONS — 14
  2.1 How do you quantify emission reductions from retrofit projects? — 14
  2.2 What are MOVES and NMIM and why should they be used to quantify emission reductions from retrofits? — 14
  2.3 How should the data included in MOVES and NMIM be used when quantifying emission reductions from retrofits? — 15
  2.4 Can you use MOVES or NMIM to estimate emission reductions from retrofit projects for uses other than SIPs or conformity determinations? — 16
  2.5 How do you use MOVES2010b to quantify emission reductions from retrofit projects? — 16
    2.5.1 Pollutant — 17
    2.5.2 Pollutant Process — 17
    2.5.3 Fuel Type and Source Type — 18
    2.5.4 Initial Calendar Year and Final Calendar Year of Retrofit Implementation — 18
    2.5.5 Initial Model and Final Model Year Retrofitted — 18
    2.5.6 Fraction of Fleet Retrofit — 18
    2.5.7 Effectiveness of the Retrofit — 19
  2.6 How do you use MOVES to quantify emission reductions from vehicle or engine replacements? — 20
  2.7 How do you use NMIM to quantify emission reductions from nonroad retrofit projects? — 21
    2.7.1 Differences between "fleet specific" and "fleet wide" retrofit projects — 21
    2.7.2 The retrofit parameters — 22
    2.7.3 The fleet information parameters — 23
  2.8 How do you use NMIM to quantify emission reductions from vehicle, engine, or equipment replacements? — 25

SECTION 3: USING EMISSION REDUCTIONS IN SIPS   27
   3.1   What are the basic requirements for using emission reductions in SIPs?   27
       3.1.1   Quantifiable -   27
       3.1.2   Surplus -   27
       3.1.3   Federally Enforceable -   28
       3.1.4   Permanent -   29
       3.1.5   Adequately Supported -   29
   3.2   How can the estimated emission reductions be used for SIP purposes?   30
   3.3   What would a state submit to EPA to meet the requirements for incorporating a retrofit project in a SIP?   31
   3.4   Are there any other types of SIPs that could include emission reductions from retrofit projects?   31
   3.5   What monitoring and record keeping should occur to document retrofit emission reductions?   32
       3.5.1   What should the state air agency monitor and record?   32
       3.5.2   How long should the state air agency maintain records?   32
   3.6   What validation and reconciliation should occur for emission reductions in SIPs approved under the VMEP guidance?   33
   3.7   What penalties can EPA impose for not complying with Clean Air Act requirements?   33

SECTION 4: USING EMISSION REDUCTIONS IN TRANSPORTATION CONFORMITY DETERMINATIONS   34
   4.1   What is transportation conformity?   34
   4.2   What kinds of retrofit projects can be used in transportation conformity determinations?   34
   4.3   How can the emission reductions from on-road retrofit projects be included in transportation conformity determinations?   35
   4.4   How can the emission reductions from nonroad retrofit projects be included in transportation conformity determinations?   36
   4.5   How is EPA supporting the use of emission reductions from nonroad retrofit projects in transportation conformity?   37
   4.6   What is a safety margin?   37
   4.7   When can a safety margin be established in a SIP?   38
   4.8   What are the benefits of establishing a safety margin at the time that an RFP SIP, attainment demonstration, or maintenance plan is initially submitted?   38
   4.9   Why should an area start developing a safety margin now?   39
   4.10   What are the features of a safety margin?   39
   4.11   What is a trading mechanism?   40
   4.12   When can a trading mechanism be established in a SIP?   41
   4.13   Why develop a trading mechanism as a stand-alone SIP revision?   41
   4.14   What would a SIP with a trading mechanism look like?   41
   4.15   What are the features of a trading mechanism?   42
   4.16   Does the public have the opportunity to comment on trading?   43
   4.17   Are there any examples of transportation conformity-related trading mechanisms that were approved into the SIP?   43
   4.18   Are there any other ways to establish a trading mechanism for a nonroad retrofit project?   44

4.19   How can I get additional technical assistance in using the emission reductions from a nonroad retrofit project in transportation conformity?   44

SECTION 5: USING EMISSION REDUCTIONS IN GENERAL CONFORMITY DETERMINATIONS   46

5.1   How can the estimated emission reductions be used for general conformity determinations?   46

5.2   How can retrofit projects be used to meet the general conformity requirements?   46

5.3   Is a SIP revision required if a source with a facility-wide emissions budget wants to implement a retrofit project at the facility?   47

5.4   What requirements would potentially limit the use of retrofit projects in the general conformity program?   48

5.5   Why are SIPs required for use of nonroad retrofits in transportation conformity but not required for general conformity?   48

APPENDIX A: 50
ESTABLISHING AND USING A SAFETY MARGIN TO ALLOW EMISSION REDUCTIONS FROM NONROAD RETROFITS TO BE USED IN TRANSPORTATION CONFORMITY DETERMINATIONS   50

A.1   What is an example of establishing and using a safety margin?   50

A.2   What is the most expeditious process for establishing a safety margin that results from a nonroad retrofit project?   51

A.3   Have any areas included safety margins in their SIPs?   53

APPENDIX B: 56
ESTABLISHING AND USING A TRADING MECHANISM TO ALLOW EMISSION REDUCTIONS FROM NONROAD RETROFITS TO BE USED IN TRANSPORTATION CONFORMITY DETERMINATIONS   56

B.1   What is an example of establishing and using a trading mechanism?   56

B.2   Does a trading mechanism have to be included in a SIP?   57

B.3   What is the most expeditious process for establishing a trading mechanism in a SIP, and how long will it take?   57

B.4   Has any state adopted a trading mechanism into its SIP this quickly?   59

APPENDIX C: 62
MODEL RULE FOR TRADING EMISSION REDUCTIONS FROM NONROAD RETROFIT PROJECTS FOR TRANSPORTATION CONFORMITY   62

SECTION 1: INTRODUCTION

1.1 What is the purpose of this guidance?

Technology is available to reduce diesel vehicle and engine emissions in a cost-effective way. The ability to use diesel emission reductions for state implementation plan (SIP) and conformity purposes gives states and localities additional incentive to implement diesel retrofit projects. Diesel retrofit technologies reduce pollution from the existing diesel engine fleet by up to 90% for particulate matter (PM), up to 75% for nitrogen oxides (NOx), and up to 90% for volatile organic compounds (VOCs). Many diesel retrofit projects are being successfully implemented around the country. Clean diesel projects already initiated are expected to result in approximately 20,000 tons of particulate matter reduced over the life of the projects, with estimated public health benefits of about $5 billion.

The purpose of this document is to provide guidance on quantifying and using emission reductions from on-road and nonroad diesel vehicles, engines, and equipment that have been retrofitted with emission reduction technology. This guidance document describes how to quantify and use reductions of NOx, VOCs, $PM_{2.5}$, PM_{10}, and carbon monoxide (CO) in ozone, $PM_{2.5}$, PM_{10}, nitrogen dioxide (NO_2), and CO nonattainment and maintenance areas. You can use the emission reductions resulting from implementing a retrofit project in a SIP to help achieve reasonable further progress (RFP), attainment of the national ambient air quality standard (NAAQS or "standard"), or maintenance of the NAAQS; and in transportation conformity and general conformity determinations. This guidance document is updated to reflect the new quantification procedures to use EPA's latest on-road emissions model, the MOtor Vehicle Emissions Simulator (MOVES) model.[1] For nonroad emissions, the nonroad portion of the National MOBILE Inventory Model (NMIM) is the preferred way to estimate benefits of retrofit programs. State and local agencies developing SIPs and conformity analyses for California should consult with EPA Region 9 for information on the current version of EMFAC approved for use in California and for information on how to quantify emission reductions from retrofit projects.

Retrofit projects provide a unique and cost-effective opportunity for state and local governments to reduce pollution from on-road and nonroad diesel vehicle and equipment fleets, and as a result, could assist areas in attaining the NAAQS. The current transportation law, Moving Ahead for Progress in the 21st Century (MAP-21), continues to provide federal funding for on-road retrofits under the Congestion Mitigation and Air Quality Improvement Program (CMAQ), and funding for diesel retrofit projects.[2] MAP-21 recognizes the importance of diesel retrofit projects and other mobile source emission reduction strategies and directs states and metropolitan planning organizations (MPOs) to give priority for use of CMAQ funds for projects that reduce fine particulate matter emissions, including diesel retrofits in $PM_{2.5}$ nonattainment and maintenance areas. The law also notes that states and MPOs continue to have final CMAQ project selection authority.

---

[1] The previous version of this document, EPA420-B-06-005, was released in June 2006. Today's version supersedes the previous document.
[2] 23 U.S.C 149(b)(8)

This guidance also fulfills the directive from the Energy Policy Act of 2005[3] that requires EPA to provide SIP guidance for retrofit projects under the diesel emission reduction provision (also known as "DERA") and takes the additional step of providing guidance for crediting retrofits in conformity determinations. In addition to assisting MPOs and states in evaluating diesel retrofits for CMAQ project selection, we anticipate that this guidance will also be useful for implementing the DERA program. DERA authorizes federal funds for retrofit projects that reduce diesel emissions from existing engines.

This guidance is focused on quantifying emission reductions from diesel retrofits for SIP and conformity purposes and therefore has an intended audience of air quality and transportation planners. Other audiences can use this guidance for quantifying emissions reductions for non-SIP or conformity purposes by reading Section 1 for background and referring to Section 2 of the guidance for quantifying emission reductions. This guidance document is written for current and future NAAQS as well as current and future versions of MOVES. EPA will re-evaluate the applicability of this guidance as needed.

1.2 *How can emission reductions from retrofit projects be used to meet SIP and conformity needs?*

This guidance is intended to facilitate the development of retrofit projects as a cost-effective way to achieve needed emission reductions while ensuring that these projects meet SIP and conformity requirements. This document describes several different options for the use of emission reductions from retrofit projects to meet both near term and longer term SIP and conformity needs.

State and local air and transportation agencies should work together to determine how reductions from diesel retrofits should be used given local air quality needs. The interagency consultation process can also be used to identify retrofit projects to implement for future SIP or conformity needs. If those reductions are used in the near term for conformity, then state and local agencies should work together to determine what additional retrofit projects could be implemented to meet any additional air quality needs.

Under any of the options highlighted below, it is in the best interests of state and local agencies to act as quickly as possible to get retrofit projects in place. Retrofit projects will provide their greatest benefits in the near term, while significant numbers of vehicles, engines, and equipment built before EPA's 2007/2010 on-road and 2008 nonroad standards remain in the fleet.

1.2.1 On-road and nonroad retrofit projects in SIPs

One option is to use retrofit reductions to help demonstrate RFP, attainment, or maintenance

---

[3] See Energy Policy Act of 2005, Title VII, Subtitle G (sections 791 to 797) at http://www.epa.gov/oust/fedlaws/epact_05.htm

in upcoming SIP submissions. On-road and nonroad retrofit projects are subject to the same SIP requirements as any other control measures. This guidance document provides the necessary information to include retrofit projects in a SIP, including a calculation method that ensures that emission reductions calculated for the retrofit project are consistent with the rest of the SIP. State and local agencies can include these retrofit projects in the SIPs being developed to meet the applicable NAAQS, but they will want to consider implementing them as soon as possible to ensure the maximum available reductions from the retrofit projects. General SIP requirements for retrofit projects are described in Section 3 of this document.

1.2.2 On-road retrofit projects in transportation conformity

Alternatively, state and local agencies could also use on-road retrofit projects to meet transportation conformity requirements now or in the future with little additional effort beyond what is required to properly implement the project and quantify the emission reductions. **These reductions could be incorporated into a transportation conformity determination without making any change in the SIP.** Section 4.3 of this document explains this process.

1.2.3 Nonroad retrofit projects in transportation conformity

This document also provides guidance on options for using reductions from nonroad retrofit projects in regional emissions analyses for transportation conformity determinations (see Section 4). **These options require a SIP revision before they can be implemented.** One option, adding a safety margin to the SIP, could be implemented in the context of the process of developing a new SIP to meet an applicable NAAQS. SIP safety margins have been implemented and used to meet transportation conformity requirements many times in the past. We have provided detailed information on how this option works in Section 4 and Appendix A.

Another option for using the nonroad retrofit reductions in transportation conformity determinations is to include a trading mechanism in the SIP. This can be done as a SIP prior to the completion of an RFP or attainment SIP. We have provided detailed information on how to implement a trading mechanism as a SIP in Section 4 and Appendix B. To help reduce the time needed to develop, submit and approve a trading mechanism SIP, we have provided a model trading rule in Appendix C. States that adopt this rule without significant revisions could expect an expedited approval by EPA.

EPA also notes that reductions from retrofits can be applied as mitigation measures in quantitative $PM_{2.5}$ or PM_{10} hot-spot analyses for project level conformity determinations. The methodology in Section 2 of this guidance can be used for such analyses. For further background on PM hot-spot analyses, see EPA's Transportation Conformity Guidance for Quantitative Hot-spot Analyses in $PM_{2.5}$ and PM_{10} Nonattainment and Maintenance Areas at www.epa.gov/otaq/stateresources/transconf/projectlevel-hotspot.htm#pm-hotspot

1.2.4 Retrofit projects in general conformity

Finally, this guidance also provides information on the use of retrofit reductions to meet general conformity requirements (see Section 5). **The use of nonroad retrofit reductions in general conformity does not require a SIP revision.**

1.3 *What vehicles, engines, and equipment does this guidance address?*

This guidance focuses on emission reductions from heavy-duty on-road and nonroad diesel vehicles, engines, and equipment, and fuels.[4] This guidance excludes locomotive and marine applications because they are not currently contained in NMIM. New emission standards, enacted in 2001, affected all on-road heavy-duty highway vehicles and engines, starting in 2007 model year, with an additional set of NOx requirements beginning in 2010. Tighter emission standards (Nonroad Tier 4) were phased in for 2008 and future model year nonroad engines. Ultra-low sulfur diesel fuel (15 ppm sulfur content) was required for on-road use beginning in October 2006. For nonroad vehicles, engines, and equipment, low sulfur diesel fuel (500 ppm sulfur content) was required beginning in 2007, with ultra-low sulfur diesel fuel required beginning in 2010. In general, this guidance applies to the retrofit of vehicles, engines, and equipment manufactured before those standards took effect and that will not have to comply with EPA's 2007/2010 on-road and 2008 nonroad regulations. This guidance document can be used, however, for emission reductions from retrofits of post-2007/2010 and post-2008 vehicles, engines, and equipment if such activities meet the definition of "retrofit" discussed below.

1.4 *What is a retrofit project?*

For the purpose of this guidance only, a "retrofit project" is defined broadly to include any technology, device, fuel, or system that when applied to an existing diesel engine achieves emission reductions beyond that currently required by EPA regulations at the time of its certification. Therefore, for those existing vehicles, engines, or equipment that will not have to comply with EPA's 2007/2010 heavy-duty on-road and 2008 nonroad standards, any additional emission reduction beyond the current regulation of these vehicle, engine, or equipment emission levels is considered a retrofit project. These technologies may include, but are not limited to, the following:

- EPA "verified" emission control technologies (for example, oxidation catalysts and PM filters and upgrade kits[5])

---

[4] On-road sources include vehicles used on roads for transportation of passengers and freight. These sources are also sometimes referred to as highway sources. Nonroad sources include vehicles, engines, and equipment used for construction, agriculture, nonroad transportation, recreation, and many other purposes. These sources are also sometimes referred to as off-road sources. Within these broad categories, on-road and nonroad sources are further distinguished by size, weight, use, and/or horsepower.

[5] For a complete list of all EPA verified technologies, consult the list at the following web site: http://www.epa.gov/cleandiesel/verification/verif-list.htm . Any diesel fuel and/or diesel fuel additive included on

- California's Air Resources Board (CARB) "verified" emission control technologies (see paragraph below)

- EPA certified engines[6] as engine replacements (see Section 1.5)

Technologies that have not been verified by EPA or CARB and operational strategies are beyond the scope of this guidance document.

In 2004, EPA signed a Memorandum of Agreement (MOA)[7] with CARB to coordinate the verification of diesel retrofit technologies. This MOA commits the agencies to establish reciprocity in verification of hardware or device-based retrofits, and further establishes our joint commitment to cooperate on the evaluation of retrofit technologies. This agreement commits EPA and CARB to accept each other's verification of the amount of PM and NOx reduced by a particular retrofit technology. Currently, EPA accepts CARB verified technologies, and CARB accepts EPA verified technologies. Additionally, as retrofit manufacturers initiate and conduct in-use testing, both agencies agreed to coordinate this testing so data generated may satisfy the requirements of both agencies. This MOA is intended to expedite the verification and introduction into the market of innovative emission reduction technologies. Additionally, this MOA reduces the effort needed for retrofit technology manufacturers to complete verification.

The verification of diesel technologies is specific to particular types of vehicles, engines, or equipment as defined in the "applicability" section of the EPA and CARB verified technology lists. Verified retrofit technologies can be applied only to the vehicles, engines, or equipment specified in those lists. For example, a technology that has been verified for on-road vehicles will not necessarily be suitable for nonroad vehicles, engines, or equipment unless it is also been verified for such use.

1.5 *What are the requirements for using vehicle, engine, or equipment replacements in SIPs and conformity?*

In addition to the EPA and CARB verified retrofit technologies, this guidance also applies to the use of EPA certified engines as engine replacements,[8] or the early replacement of older vehicles or equipment (e.g., bulldozers) with cleaner vehicles or equipment. Emission reductions that result from vehicle, engine, or equipment replacements that would have occurred through normal attrition

---

EPA's list or verified technologies is required to be registered in accordance with regulations at 40 CFR 79, Registration of Fuels and Additivies. An engine upgrade kit consists of parts (e.g., turbo charger, fuel injector, pistons, etc) that are used to rebuild an older engine to an engine configuration which is certified to meet more stringent current federal emissions standards.

[6] For a complete list of all EPA certified large on-road and nonroad engines, consult the list at the following web site: www.epa.gov/otaq/certdata.htm.

[7] The MOA is at the following web site: http://www.epa.gov/cleandiesel/documents/epa-arb_moa.pdf. The CARB verified technology list can be found here: http://www.arb.ca.gov/diesel/verdev/vt/cvt.htm.

[8] Engine replacement is sometimes referred to as an engine "repower;" EPA uses only the term "replacement" in this guidance.

are considered to be the result of normal fleet turnover and are not addressed by this guidance. The effects of fleet turnover are already included in the MOVES and NMIM/NONROAD models. Furthermore, the purchase of new vehicles or equipment to expand a fleet is not covered by the guidance. To be considered a replacement, the purchase of new vehicles, engines, and equipment would need to be accompanied by the scrappage of old vehicles, engines, and equipment and occur before normal attrition.

To be able to use emission reductions from any replacements in a SIP or in a conformity determination, the following statements would apply:

- The vehicle, engine, or equipment being replaced would be scrapped (permanently disabled or destroyed), or the replaced engine would be returned to the original engine manufacturer for remanufacturing to a cleaner standard.

- The replacement vehicle, engine, or equipment would perform the same function as the vehicle, engine, or equipment that is being replaced (e.g., an excavator used to dig pipelines would be replaced by an excavator that continues to dig pipelines).

- The replacement vehicle, engine, or equipment would be of the same type and similar gross vehicle weight rating or horsepower as the vehicle, engine, or equipment being replaced (e.g., a 300 horsepower bulldozer is replaced by a bulldozer of similar horsepower).

In addition, when emission reductions are calculated for replacements, note that:

- The emission reductions are available only for the remaining useful life of the vehicle, engine, or equipment being replaced (e.g., if the vehicle being replaced had a remaining useful life of 5 years, the additional emission reductions from the new vehicle are available for SIP or conformity purposes under this guidance only for 5 years).

- The emission reductions are available only for activity (e.g., travel or hours of use) of the vehicle, engine, or equipment being replaced within the nonattainment or maintenance area. Thus, if you replace an older, less used vehicle or piece of equipment with a piece of new equipment, it is likely that the new equipment would be used more than the item it replaces. However, emission reductions from the new equipment would need to be calculated based on the activity level of the old equipment.

1.6 How does this guidance relate to the Voluntary Mobile Source Emission Reduction Program SIP guidance?

In October 1997, EPA issued its "Guidance on Incorporating Voluntary Mobile Source Emission Reduction Programs in State Implementation Plans (SIPs)" (the VMEP guidance).[9] The

---

[9]This guidance is found at: http://www.epa.gov/otaq/stateresources/policy/general/vmep-gud.pdf EPA notes that the VMEP guidance is related to SIP measures, rather than transportation or general conformity determinations. Control

purpose of the 1997 VMEP guidance is to support innovative methods in achieving emission reductions for SIPs. The VMEP guidance applies to SIP emission reduction measures that rely on voluntary actions of individuals and other parties.

Many state or federally-funded retrofit projects may not be subject to the VMEP guidance if they have well-defined requirements to ensure the full implementation of a SIP program. For example, retrofit projects would not be subject to the VMEP guidance if a state or local regulation or ordinance that required retrofit projects was included in the SIP. Another example of a project that would not be subject to the VMEP guidance would be a state program that is described in the SIP that requires state transportation construction contracts to be implemented with retrofitted construction equipment. Please consult your EPA Region early in the development of a retrofit project to determine the appropriate use of the VMEP guidance. Retrofit projects that are pre-empted under applicable provisions of Clean Air Act section 209, as discussed in Section 1.7, may not be included in the SIP.

Under the VMEP guidance, the amount of emission reductions allowed for voluntary mobile measures in a SIP is presumed to be no greater than three percent of the total projected future year emission reductions required to attain the applicable air quality standards. EPA acknowledges that it may be possible to demonstrate that voluntary measures may achieve credible reductions higher than the three percent cap. In that case, EPA will re-evaluate that cap on a case-by-case basis and allow the cap to be exceeded if the cap hinders the implementation of effective voluntary control measures, subject to notice and comment during SIP approval. Today's guidance relies on EPA's 1997 VMEP guidance for voluntary retrofit projects. Interested parties should refer to that guidance at the time a specific retrofit project is under development.

1.7 *Can a state require retrofits of on-road and nonroad vehicles, engines, or equipment?*

The answer to this question depends on the circumstances for an individual state. Clean Air Act section 209 sets forth certain restrictions on the abilities of states and localities to adopt and implement emission standards for certain on-road and nonroad vehicles, engines, and equipment. The effect of Clean Air Act section 209 on a particular state's retrofit requirements can vary depending on the specific provisions of those retrofit requirements. State and local agencies should check with EPA before promulgating any state or local regulations or programs mandating retrofit projects for SIP or conformity purposes. See Section 1.9 of this guidance for EPA contact information.

---

measures for conformity determinations must meet the relevant criteria in the transportation and general conformity regulations.

1.8 Does this guidance create any new requirements?

This guidance does not create any new requirements, but explains to state and local air agencies, transportation agencies, MPOs, and the general public how the air quality benefits of commuter programs could be included in a SIP or in a transportation conformity determination. The Clean Air Act (CAA) and implementing regulations contain legally binding requirements. SIP requirements can be found in Clean Air Act sections 110(a)(2) and 172(c). Transportation and general conformity requirements can be found in Clean Air Act section 176(c) and applicable regulations (40 CFR Parts 51 and 93). This guidance document does not substitute for those provisions or regulations, nor is it a regulation itself. Thus, it does not impose binding, enforceable requirements on any party, and may not be applicable in all situations. EPA and State decision makers retain the discretion to adopt approaches for approval of SIP measures that differ from this guidance where appropriate and consistent with applicable law and regulations. Any final decisions by EPA regarding a particular SIP measure will only be made based on the statute and regulations in the context of EPA notice and comment rulemaking on a submitted SIP revision. This guidance may be revised periodically without public notice.

1.9 Who should I contact for additional information?

If this guidance document does not answer a specific question, please contact the appropriate EPA regional office with responsibility for air quality planning and/or conformity in the area where the retrofit project is located. A contact list of EPA Regions is available at the following web address: www.epa.gov/epahome/locate2.htm. In addition, contact information for EPA regional transportation conformity staff can be found at the following website: www.epa.gov/otaq/stateresources/transconf/contacts.htm.

For general questions regarding retrofit projects or the application of verified retrofit technologies for the existing fleet of on-road and nonroad vehicles, engines, and equipment, please contact EPA's National Clean Diesel Campaign at cleandiesel@epa.gov.

For technical questions regarding the use of MOVES or NMIM/NONROAD for calculating emission reductions from retrofit projects, please contact EPA's Office of Transportation and Air Quality at mobile@epa.gov.

For general questions concerning the use of emission reductions from retrofit projects in SIPs or in transportation conformity, please contact Astrid Larsen of EPA's Office of Transportation and Air Quality at (734) 214-4812, larsen.astrid@epa.gov or Gary Dolce also at EPA's Office of Transportation and Air Quality at (734) 214-4414, dolce.gary@epa.gov.

For general questions concerning the use of emission reductions from retrofit projects in general conformity, please contact Tom Coda of EPA's Office of Air Quality Planning and Standards at (919) 541-3037, coda.tom@epa.gov.

SECTION 2: QUANTIFYING RETROFIT EMISSION REDUCTIONS

2.1 How do you quantify emission reductions from retrofit projects?

To estimate emission reductions from retrofit projects for SIPs and for conformity analyses in states other than California, EPA recommends the use of the following models:

- The MOtor Vehicle Emissions Simulator (MOVES) for on-road vehicles[10] and
- The National Mobile Inventory Model (NMIM) containing NONROAD2008 for nonroad sources contained in the model.[11]

The most recent approved versions of these models should be used. Users should check EPA's MOVES and NMIM websites for the most current approved version of these models (MOVES2010b and NMIM2008 as of the release date of this guidance document). EPA intends to integrate nonroad capabilities into a future version of MOVES.

EPA is not providing a methodology at this time for the quantification of emission reductions from retrofit projects in California. State and local agencies developing SIPs and conformity analyses for California should consult with EPA Region 9 for information on the current version of EMFAC approved for use in California and for information on how to quantify emission reductions from retrofit projects.

2.2 What are MOVES and NMIM and why should they be used to quantify emission reductions from retrofits?

MOVES is EPA's latest motor vehicle emissions model for state and local agencies to estimate volatile organic compounds (VOCs), nitrogen oxides (NOx), particulate matter (PM), carbon monoxide (CO), and other precursors from cars, trucks, buses, and motorcycles for SIP purposes and conformity determinations outside of California. MOVES is available at http://www.epa.gov/otaq/models/moves/index.htm.[12] MOVES replaced MOBILE6.2, EPA's previous emissions model. MOVES is required for SIPs begun after December 2009 and regional transportation conformity analyses begun after March 2, 2013. See EPA's "Policy Guidance on the

---

[10] This guidance is applicable to current and future versions of the MOVES model, unless EPA notes otherwise when approving the model for conformity purposes. Modelers should follow the "Using MOVES to Prepare Emission Inventories in State Implementation Plans and Transportation Conformity: Technical Guidance for MOVES2010, 2010a, and 2010b," EPA-420-B-12-028 (April 2012); available online at www.epa.gov/otaq/models/moves/index.htm#sip. The MOVES model, user guide, and supporting documentation are available online at www.epa.gov/otaq/models/moves/index.htm.

[11] NMIM2008 and supporting documentation are available online at www.epa.gov/otaq/nmim.htm. NONROAD2008 could also be used, but EPA does not recommend this because NONROAD2008 does not include the fleet specific capabilities included in NMIM. NONROAD2008 also may not work with newer 64-bit operating systems.

[12] Note that this guidance refers to "MOVES" generally rather than a particular MOVES version because EPA will release updated versions of MOVES in the future.

Use of MOVES2010 and Subsequent Minor Revisions for State Implementation Plan Development, Transportation Conformity, and Other Purposes" (available at www.epa.gov/otaq/models/moves/documents/420b12010.pdf) for more details. MOVES includes features that are specifically intended for modeling retrofit programs which are described in detail in Sections 2.5 and 2.6 below.

NMIM is a graphical user interface (GUI) that contains the NONROAD2008 and (the now no longer used) MOBILE6.2 models and a database of county-level input information, called the National County Database (NCD). NMIM was created to simplify the process of developing county-by-county emissions inventories for multi-county areas, states, or the entire nation. MOVES has replaced NMIM for on-road emissions estimation and, as a result, the use of on-road emissions estimates from NMIM is not currently acceptable for any regulatory purpose. However, the portion of NMIM that contains NONROAD2008 is the best option for modeling retrofit, rebuild, and replacement programs that affect nonroad equipment. The use of NMIM to model retrofit programs for nonroad equipment is described in detail in Sections 2.7 and 2.8 below. NMIM is available at www.epa.gov/otaq/nmim.htm.

MOVES and NMIM incorporate EPA's latest emissions data for on-road and nonroad sources respectively, as well as the capabilities to apply retrofit benefits to emission factors or inventories that are generated by MOVES and NONROAD under local conditions. The emissions of retrofit highway and nonroad vehicles, engines, and equipment are subject to the same external factors as are the emissions for all other vehicles, engines, and equipment. These external factors include environmental factors (e.g., temperature, humidity), fleet characteristics (e.g., age distribution of fleet, distribution of VMT by vehicle class, number and types of nonroad engines or equipment), activity measures (e.g., speed distributions, distribution of VMT by roadway type, distribution of hours of operation for nonroad equipment), and fuel characteristics (e.g., sulfur content, Reid vapor pressure (RVP)). The majority of impacts of these external factors on the emissions of vehicles, engines, and equipment meeting past, current, and future emission standards are incorporated in EPA's MOVES and NONROAD emissions models which are used to develop emissions inventories for SIPs and for regional transportation conformity analyses. Using MOVES and NMIM to generate retrofit emission reductions ensures that those reductions are based on the same conditions used to generate the rest of the inventory used in the SIP or conformity analysis.

2.3 *How should the data included in MOVES and NMIM be used when quantifying emission reductions from retrofits?*

As with any model, the quality of the inputs in MOVES and NMIM affects the quality of the model results. As required by Clean Air Act Section 172(c)(3) and EPA's regulations at 40 CFR 51.112(a), States must use the latest planning assumptions available at the time the SIP is developed, including, but not limited to VMT, speeds, fleet mix and SIP control measures.

MOVES includes a national default database of fleet and activity information that is applied at the county level. However, in most cases, local data should be used instead of default fleet and activity data. Detailed guidance on the development of MOVES inputs for SIP and conformity analyses is included in "Using MOVES to Prepare Emission Inventories in State Implementation Plans and Transportation Conformity: Technical Guidance for MOVES2010, 2010a, and 2010b" (hereafter referred to as the MOVES Technical Guidance) available on the MOVES web page at http://www.epa.gov/otaq/models/moves/index.htm#sip. All inputs used in estimating retrofit benefits for SIP or conformity purposes should be consistent with the MOVES Technical Guidance. This retrofit guidance document addresses the specific inputs needed for on-road retrofit programs which are not covered in the MOVES Technical Guidance.

As stated above, NMIM should be used only for modeling retrofit, rebuild, and replacement programs that affect nonroad equipment. NMIM includes a database of county-level nonroad information called the National County Database (NCD). States have provided some local information for the National County Database (NCD) as part of the National Emissions Inventory (NEI) development process. However, given the NEI cycle, this may not be the most recent or best available information at the time a state initiates modeling. For SIPs and regional conformity analyses, state and local agencies should review the information in the NCD to evaluate whether it includes the latest and best information currently available. Where more current or better information is available, the database must be modified to incorporate it to meet regulatory requirements for the use of latest planning assumptions in SIPs and conformity determinations.[13] The interagency consultation process should be used to evaluate what changes are needed in the NMIM database for the area.

2.4 Can you use MOVES or NMIM to estimate emission reductions from retrofit projects for uses other than SIPs or conformity determinations?

Yes. MOVES could be used to evaluate on-road retrofit projects and NMIM could be used to evaluate nonroad retrofit projects for other purposes, such as the development of proposals for retrofit projects. For these purposes, MOVES and NMIM users could rely more on default data in MOVES and NMIM or other more simplified methods for using MOVES and NMIM than would otherwise be necessary for SIP or conformity purposes. These simplified methods may result in emission reduction estimates that are not completely consistent with emission reductions calculated for SIP or conformity purposes using more rigorous methods. Consultation between organizations developing project proposals and state and local air quality and transportation agencies about appropriate methods and interpretation of results is important to ensure that retrofit projects are properly evaluated.

2.5 How do you use MOVES2010b to quantify emission reductions from retrofit projects?

Retrofit projects are modeled using the On-Road Retrofit option in the Control Strategies panel to create an on-road retrofit input file in MOVES. The details of how to do this in MOVES2010b are described in Section 2.2.9 and Appendix D of the MOVES2010b User Guide

---

[13] EPA encourages states to separately submit updates to the NCD so that the most accurate database is available for both national and local inventory development.

(available at http://www.epa.gov/otaq/models/moves/documents/420b12001b.pdf.) This part of this guidance document summarizes some of the key inputs for MOVES and discusses some of the issues that users face when developing input data. Note that, unlike all other MOVES input files, the retrofit parameter file uses abbreviated names rather than ID numbers for most of the input parameters.

EPA has supplied a Retrofit Converter tool for simplifying the creation of an on-road retrofit input file, which is available at www.epa.gov/otaq/models/moves/tools.htm under the heading "Retrofit Converter". The Retrofit Converter allows the user to describe the retrofit project in terms of the verified technology and vehicles to which it applies (type, number, model year), and creates the appropriate input file for MOVES. The tool automatically accounts for chained pollutants (i.e., emissions for pollutants that are dependent in MOVES on the emissions of another pollutant), creates input lines for each pollutant process, and uses the appropriate abbreviated name, among other things. For SIP and conformity purposes, the Retrofit Converter tool should be used in a manner consistent with the guidance in this document and the MOVES Technical Guidance. The instructions that come with the tool do not provide all of the detail included in this guidance and are not a substitute for it. The Retrofit Converter does have some limitations which are described in Sections 2.5.2 and 2.5.7. Users of the Retrofit Converter should always check the output of the converter to make sure it is consistent with the rest of this guidance.

The MOVES retrofit input file describes the details of the retrofit project. This includes inputs that specify the pollutants and pollutant processes affected by the retrofit project, the fuel types affected, the vehicle types affected, the calendar years during which the retrofits occur, the model years of the vehicles that will be retrofitted, the percentage of the fleet VMT accounted for by retrofit vehicles per year, and the percentage effectiveness of the retrofit technology applied to the vehicles. Details on the use of these inputs are described in the Appendix D of the MOVES User Guide. Specific guidance on these inputs, where applicable, is given below.

2.5.1 Pollutant

EPA and CARB have verified emission reductions from retrofit technologies only for NOx, VOC, CO, PM_{10}, and $PM_{2.5}$ emissions. Therefore, when estimating emission reductions from retrofit projects for a SIP or conformity purposes, only these pollutants and any pollutants that are needed by MOVES to calculate emissions of these pollutants should be selected (see Section 2.2.7.2 of the MOVES User Guide for a list of base and dependent pollutants or use the Retrofit Converter tool to automatically include any other pollutants are needed to calculate the emission of the pollutant you are estimating). As discussed in Section 2.5.7 below, a retrofit evaluation for SIP or conformity purposes should include all pollutants for which the project area is nonattainment or maintenance, including any that increase as a result of the retrofit technology.

2.5.2 Pollutant Process

The retrofit parameter file will include a separate line for each pollutant process affected by the retrofit technology. All retrofit parameter files must include start and running exhaust processes for each pollutant for which you are estimating reductions. Some retrofit technologies may also reduce crankcase running and crankcase start emissions and will include lines for each of these processes. There are no reductions associated with the extended idle process for retrofit technologies included in the EPA and CARB verified technology lists. The Retrofit Converter tool automatically creates the inputs for start and running exhaust processes, but not for crankcase start and running exhaust processes, which must be entered separately. Section 2.5.7 of this document provides additional guidance on selecting pollutant process inputs for technologies that reduce crankcase emissions.

2.5.3 Fuel Type and Source Type

As discussed in Section 1.3, this guidance document addresses on-road retrofit projects only for heavy-duty diesel vehicles. The retrofit input in MOVES allows you to enter the full range of on-road source types (vehicle types) and fuel types, but only inputs for heavy-duty diesel vehicles should be used for retrofit emission reductions in SIPs or conformity determinations. The Retrofit Converter will convert fuel and source type names to the appropriate abbreviated names.

2.5.4 Initial Calendar Year and Final Calendar Year of Retrofit Implementation

The Initial Calendar Year and the Final Calendar Year of Retrofit Implementation should be equal to the calendar year you are modeling. For example, if the calendar year of your analysis is 2012, the initial and final years of implementation should also be 2012. Using other years may cause an incorrect result.

2.5.5 Initial Model and Final Model Year Retrofitted

This input allows you to define the vehicle model years covered by the retrofit project. Note that the Retrofit Converter only has a single input line labeled "Vehicle Model Year". For projects that cover multiple model years, simply create a new line in the converter for each year. Using this approach will ensure that differences in fractions of the fleet retrofit by model year are accounted for in the retrofit calculation.

2.5.6 Fraction of Fleet Retrofit

This input represents the fraction of activity (VMT for running emissions and vehicle population for start emissions) associated with retrofitted vehicles for a particular model year and source type. This number can be a value from 0.00 through 1.00. For example, if retrofitted 2010 model year school buses account for 40% of all 2010 model year school bus VMT in the calendar year of analysis, you would enter 0.40 for the fraction of the school bus fleet retrofitted for the 2010 model year.

2.5.7 Effectiveness of the Retrofit

This input is used to describe the effectiveness of the particular vehicle, engine, or equipment technologies being used in the retrofit project. As mentioned in Section 1.4, EPA has verified the emission reductions for certain retrofit technologies. A list of these EPA-verified retrofit technologies, the vehicle types and model years they apply to, and the emission reductions associated with them can be found at http://www.epa.gov/cleandiesel/verification/verif-list.htm. Retrofit technologies that have been verified by the California Air Resources Board (CARB) can be found at: http://www.arb.ca.gov/diesel/verdev/vt/cvt.htm. You should use retrofit reductions from these verified retrofit technology lists as inputs for MOVES. These reductions are entered in MOVES as a fraction less than or equal to 1.00 (e.g., a 50% reduction is entered as 0.50). The same value should be added for both running and start emissions processes.

Retrofit technologies that include closed crankcase systems will require some adjustments to the values included given in the verified technology list to account for the way MOVES separates crankcase start and running emissions as separate processes from tailpipe start and running emissions. For closed crankcase technologies, add additional lines to the input file for crankcase start emissions and crankcase running emissions with an effectiveness fraction of 1.00 (i.e., 100% reduction in crankcase emissions) for PM emissions (do not add lines for crankcase emissions for any other pollutants). In addition, the values for tailpipe start and running emissions benefits for PM should be adjusted downward by 5%. For example, if the retrofit technology includes a closed crankcase system and the verified technology list gives a PM reduction of 30%, you should have four lines in the input file for PM:
1. Crankcase start with an effectiveness value of 1.00
2. Crankcase running with an effectiveness value of 1.00
3. Start exhaust with an effectiveness value of 0.25
4. Running exhaust with an effectiveness value of 0.25

The Retrofit Converter includes information from EPA's verified technology list in Column J of the User Input tab under the heading "Default Removal Efficiency" based on the technology selected in Column A, "Verified Technology." However, this information may change and the Converter may not be up-to-date. You should always look up the benefit in the verified technology list and enter that number in Column C, "User Specified Removal Efficiency." If you are relying on a technology from CARB's verified technology list, you will have to look up those reductions on the CARB website and enter them in Column C.

Note that the Converter only enters lines for the start exhaust and running exhaust processes. You will have to manually enter lines for PM for crankcase start and running emissions for technologies that include closed crankcase systems. The Retrofit Converter also does not automatically adjust the start and running exhaust emissions downward as described above. The user specified removal efficiency will have to be adjusted to reflect the decreased benefits for start and running exhaust.

Note that the emission reductions for each verified technology in the EPA and CARB verified technology lists are specific to certain categories of vehicles or engines and to certain model years. The emission reductions in those lists should be applied only to the categories of vehicle or engines and model years for which they have been specifically verified. Also note that the Retrofit Converter does not check that the vehicle types and model years given in the verified technology list for a particular retrofit technology are consistent with the vehicle types and model years entered by the user in the Converter. It is your responsibility to check and edit the input file created by the Retrofit Converter as needed to ensure it is consistent with the information in the verified technology list and this guidance.

Some verified retrofit technologies may result in emission reductions for one pollutant and emissions increases for another. Any analysis of retrofit projects for SIPs or conformity purposes should include all pollutants for which the project area is nonattainment or maintenance that are affected by the retrofit project, including any that increase as a result of the retrofit technology used. An emissions increase from a retrofit is entered in MOVES as a negative number (e.g., a retrofit that results in a 50% increase in emissions would be entered as -0.50).

For most types of retrofit projects with the exception of replacements (discussed in Section 2.6), the emission reductions for the project are best determined by using two MOVES runs - one with the percentage effectiveness of the retrofit set to zero (to represent the inventory without the project) and one with it set to the appropriate level for the technology used (to represent the inventory with the project). All other parameters in this section would be set identically to describe the affected fleet of vehicles or equipment. The difference in emissions between these two runs is the emission reduction associated with the retrofit project.

2.6 *How do you use MOVES to quantify emission reductions from vehicle or engine replacements?*

Emission reductions from vehicle or engine replacements should not be modeled using the On-Road Retrofit Strategy (or the retrofit input file) in MOVES. Instead, you will need to do two runs, using the "Fueltype and Technologies" importer in the County Data Manager to define a base case fleet and a control case fleet consisting of different model years. Use of the Fueltype and Technologies importer is discussed in the MOVES User Guide and in Section 4.9 of the MOVES Technical Guidance.

In this case, we strongly recommend that you set up the base case and control case MOVES runs to include only the fleet of vehicles affected in the project. For example, if you are evaluating a retrofit project in 2012 that involved replacing 2005 model year school buses with 2010 model year buses, emission reductions should be calculated in the following way:

1. Set up a base case RunSpec for 2012, selecting only diesel school buses in the Vehicles/Equipment panel.
2. Create the base case input files for the CDM that describe the expected activity of 2005 model year school bus in the analysis year, 2012.
3. Create an age distribution input file for the CDM which only includes 2005 model year school buses.
4. Create a Fueltype and Technologies input file in which all the school buses are 2005 model year diesel buses.

5. Run MOVES to get the emissions of 2005 model year school buses in 2012.
6. Set up a control case RunSpec identical to the base case RunSpec.
7. Use the same input files as in step 2; i.e., assume that the control case buses have the same activity as the base case buses.
8. Create a Fueltype and Technologies input file in which all the school buses are 2010 model year diesel buses.
9. Run MOVES to get the emissions of 2010 model year school buses in 2012.
10. Take the difference between the base case and control case runs to get the benefits of the replacement project.

Note that these reductions should not be used beyond the useful life of the vehicles, engines, or equipment being replaced. In this example, if the useful life of these school buses is 10 years, emission reductions could be applied under this guidance for calendar years 2010 through 2015. However, after 2015, these reductions would be considered part of the normal fleet turnover for these vehicles and would not be available for use in a SIP or in a conformity determination. See Section 1.5 for further information.

2.7 How do you use NMIM to quantify emission reductions from nonroad retrofit projects?

The details of how to use NMIM to estimate the emissions impact of nonroad retrofit projects are described in the NMIM User Guide (available at www.epa.gov/otaq/nmim.htm). This part of this guidance document summarizes some of the key inputs for NMIM and discusses some of the issues that users face when developing input data.

2.7.1 Differences between "fleet specific" and "fleet wide" retrofit projects

NMIM divides retrofit projects into two different categories: "fleet specific" and "fleet wide." A "fleet specific" retrofit project refers to those projects where a well-defined group of nonroad vehicles or engines are the targets for retrofit. A construction company implementing a retrofit project would be an example of a fleet specific project. A fleet specific project could include multiple model years, or multiple equipment types. The key defining characteristic of a fleet specific project is that the actual number of nonroad vehicles or engines, as well as their type, model year, and activity is known.

A "fleet wide" retrofit project refers to situations where the actual nonroad vehicles or engines that will be affected by the retrofit project are not known in advance. One example of a fleet wide retrofit project would be the availability of a low emission diesel fuel that applies to the entire nonattainment area rather than to specific fleets in the nonattainment area. In this case, the actual number of nonroad vehicles or engines, or their type, model year, or activity is not precisely known.

There are important differences between fleet specific and fleet wide projects that affect the kind of information that is needed to run NMIM. It is assumed that for fleet specific projects the precise number of nonroad vehicles or engines in each model year of each class that are to be retrofit will be known. In addition, it is assumed that the annual average hours accumulated by each model year of each class is also known. In general, retrofit projects that involve the addition of specific equipment to nonroad engines or vehicles, or involve the replacement of nonroad engines, vehicles or equipment, should be modeled as fleet specific projects because the entities involved with this type of project should be able to keep the records of the nonroad vehicles or engines modified or replaced and their average use.

In contrast, fleet wide projects are expected to have no precise information about the individual nonroad vehicles or engines that will be retrofitted. In this case, the NMIM model assumes that the average hours accumulated by retrofit vehicles or engines is the same as for all nonroad vehicles or engines of that model year and vehicle or engine class. Implementation is expressed as a fraction of all nonroad vehicles or engines of that model year and vehicle class. In general, retrofit projects based on fuels available to the general fleet will likely need to be modeled using the fleet wide approach.

As discussed in detail in the NMIM User Guide, the specifics of nonroad retrofit projects are described in an input file called the "Nonroad Retrofit Parameters File." This file is used for both fleet specific and fleet wide retrofit projects. For nonroad fleet specific projects, an additional input file called the "Nonroad Fleet Information Parameters File" is required. This file describes in detail the specific fleet affected by the retrofit project.

2.7.2 The retrofit parameters

The file for nonroad retrofit parameters describes the details of the retrofit project. This includes inputs that specify the pollutants affected by the retrofit project, the nonroad vehicle or engine types affected, the calendar years during which the retrofits occur, the model years of the vehicles or engines that will be retrofitted, the percentage of the fleet retrofit per year, and the percentage effectiveness of the retrofit technology applied to the vehicles or engines. Details on the use of these inputs are described in the NMIM User Guide. Specific guidance on these inputs, where applicable, is given below.

2.7.2.1 Pollutants affected by the retrofit project

While NMIM allows you to enter the entire range of pollutants for which NONROAD2008 provides emissions estimates, EPA and CARB have verified emission reductions from retrofit technologies only for NOx, VOC, CO, PM_{10}, and $PM_{2.5}$ emissions. Therefore, when estimating emission reductions from nonroad retrofit projects for a SIP or conformity determination, only these pollutants should be evaluated. As discussed in Section 2.6.2.3 below, all pollutants for which the project area is nonattainment or maintenance that are affected by the retrofit project, including any that increase as a result of the retrofit technology, should be evaluated for SIP or conformity purposes.

2.7.2.2 Nonroad vehicle or engine types affected

As discussed in Section 1.3 above, this guidance document addresses nonroad retrofit projects only for diesel nonroad vehicles, engines, and equipment. NMIM allows you to enter the full range of vehicle and engine types included NONROAD2008, but only inputs for nonroad diesel vehicles or engines should be used for retrofit emission reductions in SIPs or conformity determinations.

2.7.2.3 Percentage effectiveness of the retrofit

This input is used to describe the effectiveness of the particular vehicle, engine, or equipment retrofit technologies being used in the retrofit project. As mentioned in Section 1.4, EPA has verified the emission reductions for certain retrofit technologies. A list of these EPA-verified retrofit technologies and the emission reductions associated with them can be found at www.epa.gov/cleandiesel/verification/verif-list.htm. A link to retrofit technologies that have been verified by CARB can be found at http://www.arb.ca.gov/diesel/verdev/vt/cvt.htm. Retrofit reductions from these verified retrofit technology lists should be used as inputs for NMIM.

Note that the emission reductions for each verified technology in the EPA and CARB verified technology lists are specific to certain categories of vehicles or engines. The emission reductions in those lists should be applied only to the categories of vehicle or engines for which they have been specifically verified.

Some verified retrofit technologies may result in emission reductions for one pollutant and emissions increases for another. Any analysis of retrofit projects for SIPs or conformity purposes should include all pollutants for which the project area is nonattainment or maintenance that are affected by the retrofit project, including any that increase as a result of the retrofit technology used.

For most types of retrofit projects with the exception of replacements (discussed in Section 2.8), the emission reductions for the project are best determined by using two NMIM runs - one with the percentage effectiveness of the retrofit set to zero (to represent the inventory without the project) and one with it set to that appropriate level for the technology used (to represent the inventory with the project). All other parameters in this section would be set identically to describe the affected fleet of vehicles or equipment. The difference in emissions between these two runs is the emission reduction associated with the retrofit project.

2.7.3 The fleet information parameters

The file for nonroad fleet information is used to provide details of specific fleets of vehicles or engines for which more detailed information is known. Use this file, along with the retrofit parameter file described above, when quantifying the emission reductions from fleet specific retrofit projects. For nonroad equipment, this file includes three inputs needed to specify the engines in the project – source category classification (SCC) code, horsepower bin, and technology type – as well as model year, number of engines, activity in hours per year, and monthly activity allocation.

Details on the use of these inputs are described in the NMIM User Guide. Specific guidance on these inputs, where applicable, is given below. If you are modeling a fleet specific project, you should be able to enter detailed information for all of these inputs. Note that these files only describe characteristics of a fleet of vehicles or engines; they do not describe any details of a retrofit project. When quantifying the reductions for a fleet specific retrofit project, the fleet information parameters files are used to describe the specific fleet, while the retrofit parameter files are used to describe the retrofit project applied to that fleet. When used without retrofit parameter files, the fleet information files can be used to simply quantify the emissions for any specific fleet of vehicles or engines.

2.7.3.1 Vehicle class or specific engine parameters

Because this function can be used to quantify the emissions from any specific fleet of vehicles or engines, there is no restriction on the types of vehicles or engines that can be entered here. However, retrofit emission reductions will be applied only to those vehicles or engines specified in the retrofit parameters file, which, as described above, should include only diesel vehicles and engines.

2.7.3.2 Number of vehicles or engines

The number of vehicles or engines entered for the specific fleet should be based on the calendar year for which emission estimates are being calculated. When estimating emissions for a specific fleet in the current year, this is the current size of the fleet. However, in future years the fleet of affected vehicles or engines may become smaller as some vehicles or engines in the fleet are scrapped while other newer, non-retrofit vehicles may be added to the fleet. NMIM includes the effects of normal attrition when projecting future emissions for the entire fleet (e.g., the model assumes that the number of 1998 model year vehicles or engines decreases in each future year). However, these effects are not applied to the number of vehicles or engines entered in the fleet information file for a specific fleet (e.g., if your input file indicates that you have twenty 1998 model year vehicles or engines in your retrofit fleet in 2005, NMIM will assume twenty 1998 model year vehicles or engines in any future year that you model). If you have reason to believe that some of the vehicles currently in the specific fleet may no longer be in the fleet by the calendar year that is being evaluated, reduce the input for number of vehicles or engines appropriately.

2.7.3.3 Hours of use

The activity level (hours of use) entered for the specific fleet should be based on the activity that actually occurs within the nonattainment or maintenance area that the SIP or conformity analysis applies to. For example, in the case of a retrofit project applied to a fleet of construction equipment, you must not include hours of use when that equipment is taken outside the nonattainment or maintenance area.

The activity level (hours of use) entered for the specific fleet should be based on the calendar year for which emission estimates are being calculated. When estimating emissions for a specific fleet in a current year, this is the current activity level of the fleet. However, in future years, the activity level of the affected vehicles or engines in the fleet may change as older vehicles and engines are often used less than newer ones. NMIM includes the effects of decreased activity with age when projecting future emissions for the entire fleet (e.g., NMIM assumes that the activity of 1998 model year vehicles or engines decreases in each future year). However, these effects are not applied to the activity levels entered in the fleet information file for a specific fleet (e.g., if your input file indicates that 1998 model year tractors are used 1000 hours in 2005, NMIM will assume that 1998 model year tractors are used 1000 hours in any future year that you model). If you have reason to believe that activity levels of vehicles or engines currently in the specific fleet may be lower by the calendar year that is being evaluated, this lowered activity should be accounted for by reducing the input for hours of use appropriately.

Specific information on the hours of use of retrofit vehicles or engines may be available from maintenance records, user logs, or fuel records. In the absence of this kind of information, the interagency consultation process should be used to determine the best available information to account for activity in the calculation of emission reductions from a retrofit project. In the absence of better information, agencies could agree to use local average estimates of vehicle or equipment activity for the class and model year of vehicles or engines included in the retrofit project.

2.8 How do you use NMIM to quantify emission reductions from vehicle, engine, or equipment replacements?

In general, emission reductions from vehicle, engine, or equipment replacements should be modeled using the fleet specific approach described in Section 2.7, with some modifications. Vehicle, engine, or equipment replacement projects will need to run NMIM twice. NMIM should be run once as a base case without the replacement data and then again with the control case that accounts for the new replacement information. For a retrofit project that uses replacements, the percentage effectiveness of the retrofit should be set to zero in both the base case (without the retrofit project) and the control case (with the retrofit project). Instead, the model year of the vehicles or engines affected are varied in the fleet information parameters. For example, if the retrofit project involved replacing 2005 model year excavators with 2010 model year excavators, emission reductions would be calculated as the difference between the following two NMIM runs:

- Base case – Engine model year[14] set to 2005; all other parameters, including activity, set to describe how the model year 2005 excavators are expected to be used in the analysis year; retrofit effectiveness set to zero.

- Control case – Engine model year set to 2010; all other parameters, including activity, set to the same values as in the base case; retrofit effectiveness set to zero.

---

[14] Engine model year and chassis model year can in some cases be different.

Note that these reductions should not be used beyond the useful life of the vehicles, engines, or equipment being replaced. In this example, if the useful life of these excavators is 10 years, emission reductions could be applied under this guidance for calendar years 2010 through 2015. However, after 2015, these reductions would be considered part of the normal fleet turnover for these vehicles and would not be available for use in a SIP or in a conformity determination. See Section 1.5 for further information.

SECTION 3: USING EMISSION REDUCTIONS IN SIPS

3.1 What are the basic requirements for using emission reductions in SIPs?

In order to be approved as a control measure which provides additional emission reductions in a SIP, a retrofit project would need to be consistent with SIP RFP, attainment, or maintenance requirements and other requirements of the Clean Air Act, as appropriate. The retrofit project must provide emission reductions that meet the basic SIP requirements described below. Information is separated into "SIP Requirement" and "Specific Recommendations". The "SIP Requirement" heading refers to mandatory requirements under Clean Air Act section 110. The "Specific Recommendations" headings include our recommendations for implementing a retrofit project. While these recommendations are not binding, they may provide appropriate safeguards and considerations for a successful retrofit project.

3.1.1 Quantifiable -

SIP Requirement: The emission reductions from a retrofit project are quantifiable if they are measured in a reliable manner and can be replicated (e.g., the assumptions, methods, and results used to quantify emission reductions can be understood). Emission reductions must be calculated for the time period during which the reductions will occur and will be used for SIP purposes.

Specific Recommendations:

- In general, if you are retrofitting certified vehicles, engines, or equipment with EPA or CARB verified emission control technologies or certified engine configurations, quantifying the emission reductions is fairly straightforward. In these circumstances, you will need to document the emission reductions and provide all relevant data to EPA for review.

Section 2 of this document provides you with a recommended method for quantifying emission reductions.

3.1.2 Surplus -

SIP Requirement: Emission reductions are considered "surplus" if they are not otherwise relied on to meet other applicable air quality attainment or maintenance requirements for that particular NAAQS pollutant (i.e., there can be no double-counting of emission reductions). In the event that the retrofit project is used to meet such air quality related program requirements, they are no longer surplus and may not be used as additional emission reductions. Emissions from the vehicles, engines, or equipment to be retrofitted must be in the applicable mobile source emissions inventory before the emission reductions from a retrofit project can be used for RFP, attainment or maintenance in a SIP.

3.1.3 Federally Enforceable -

SIP Requirement: A SIP retrofit project must be enforceable. Depending on how the emission reductions are to be used, control measures must be enforceable through a SIP. Where the emission reductions are part of a rule or regulation for SIP purposes, they are considered federally enforceable only if they meet all of the following criteria:

- They are independently verifiable.
- Violations are defined, as appropriate.
- You and EPA have the ability to enforce the measure if violations occur.
- Those liable for violations can be identified.
- Citizens have access to all the emissions-related information obtained from the responsible party.
- Citizens can file lawsuits against the responsible party for violations.
- Violations are practicably enforceable in accordance with EPA guidance on practicable enforceability.
- A complete schedule to implement and enforce the measure has been adopted by the implementing agency or agencies.

The specific requirements for enforceability vary when submitting a SIP retrofit project as a mandatory[15] or voluntary measure. If your retrofit project is mandatory, then there is no cap on the amount of emission reductions that can be claimed as long as such reductions are supported and meet standard SIP enforceability requirements for mandatory measures and the baseline emissions are in the inventory.

If a retrofit control measure is approved under EPA's VMEP guidance, the state is responsible for assuring that the reductions quantified in the SIP occur. The state would need to make an enforceable SIP commitment to monitor, assess, and report on the emission reductions resulting from the voluntary measure and to remedy any shortfalls from forecasted emission reductions in a timely manner. Under the VMEP guidance, the amount of emission reductions allowed for voluntary mobile source measures, including commuter programs, in a SIP is not expected to exceed three percent of the total reductions needed to meet any requirements for reasonable further progress, attainment or maintenance, as applicable.[16] EPA acknowledges that some areas may be able to demonstrate that voluntary measures may achieve credible reductions higher than the three percent cap provided by the VMEP guidance. In that case, EPA will re-evaluate that cap on a case-by-case basis and allow the cap to be exceeded if the cap hinders the implementation of effective voluntary control measures, subject to notice and comment during SIP approval. If you wish to have a retrofit project approved as a voluntary measure, consult the 1997

---

[15] As stated in Section 1.7 of this guidance, state and local agencies should check with EPA before promulgating any state or local regulations or programs mandating retrofit projects, due to potential issues related to Clean Air Act section 209 requirements.

[16] EPA acknowledges that it may be possible to demonstrate that voluntary measures may achieve credible reductions higher than the three percent cap. In that case, EPA will re-evaluate that cap on a case-by-case basis and allow the cap to be exceeded if the cap hinders the implementation of effective voluntary control measures, subject to notice and comment during SIP approval. Interested parties should refer to the VMEP guidance for more information when a specific commuter program is under development.

VMEP guidance for further information.

3.1.4 Permanent -

SIP Requirement: The emission reduction produced by the retrofit project must be permanent throughout the time period that the reduction is used in the applicable SIP. The time period that the emission reductions from retrofit projects are used in the SIP can be no longer than the remaining useful life of the retrofitted or replaced engine, vehicle, or equipment.

Specific Recommendations:

- Emission reductions can be used from retrofitted vehicles, engines, and equipment that operate exclusively within the nonattainment or maintenance area. Vehicles, engines, and equipment that typically operate within a captive area may include, but are not limited to, the following:

    - School buses
    - Transit buses
    - Waste haulers
    - State/local government owned vehicles and engines (e.g., department of transportation)
    - Nonroad construction and agricultural vehicles, engines, or equipment

  Some fleets may travel or some equipment may be used only partially in the nonattainment or maintenance area. Such fleets may be considered as part of a retrofit project for these purposes, but the emission reductions claimed are limited to the activity (and the associated emission reductions) that are expected to occur from such fleets within the nonattainment or maintenance area.

  For regulatory or voluntary retrofit projects, you should demonstrate that the retrofitted vehicles, engines, or equipment remain in use within the nonattainment or maintenance area or their remaining useful life, to the extent emission reductions are claimed.

- EPA and CARB provide information on the durability of the verified retrofit technology which allows you to determine the length of time the technology may perform at its verified emission reduction capability. Consequently, you should select retrofit technologies that are verified or certified by EPA, or CARB respectively. For a list of verified technologies, see http://www.epa.gov/cleandiesel/verification/verif-list.htm

3.1.5 Adequately Supported -

SIP Requirement: The state must demonstrate that it has adequate funding, personnel, implementation authority, and other resources to implement the retrofit project on schedule.

Specific Recommendations:

- The state should ensure it has allocated appropriate funds from a reliable funding source.

- The state should ensure that the retrofit fleet operators correctly install, operate, and maintain the retrofit technology according to the manufacturer's recommendations.

    o <u>Example</u>: The city transit fleet has 50 buses retrofitted with PM filters. The state should ensure that the fleet operators are properly trained to operate, maintain, and detect problems with the PM filters.

- The state should assess and verify the status of the retrofitted vehicles, engines, and/or equipment and the associated emission reductions, as applicable.

3.2 How can the estimated emission reductions be used for SIP purposes?

For your RFP, attainment, or maintenance SIP strategy, you can use emission reductions which are expected to be generated from the retrofit project by applying the following criteria:

- Emission reductions would be calculated as required in the SIP process for a given pollutant and NAAQS, either in tons per year, tons per pollutant season, or tons per day. For example, NOx reductions from retrofit projects would be calculated in an ozone SIP for tons reduced per day for a typical summer day. In contrast, $PM_{2.5}$ reductions would be calculated on a tons per year basis for SIP inventories for the annual $PM_{2.5}$ NAAQS; state and local agencies should consult their EPA Region and/or applicable guidance on what is appropriate for SIP inventories for the 24-hour $PM_{2.5}$ NAAQS. Any calculations would consider factors that may affect emission reductions and their surplus status over time, including changing patterns of operations or use, vehicle deterioration factors, equipment useful life, and government emission standards.

- Emission reductions would be commensurate with the level of activity from retrofitted vehicles, engines, or equipment within a given nonattainment or maintenance area as described in Section 3.1.4. For example, if retrofitted vehicles are operated exclusively within the nonattainment or maintenance area, the associated reductions from retrofit technology would also be assumed to occur within such an area. However, some fleets may leave the nonattainment or maintenance area for some portion of their operation. Such fleets may be considered as part of a retrofit project, but the emission reductions claimed are limited to the activity that is expected to occur from such vehicles within the nonattainment or maintenance area, as well as those accounted for in the inventory.

3.3 What would a state submit to EPA to meet the requirements for incorporating a retrofit project in a SIP?

A state would submit to EPA a written document which does the following:

- Identifies and describes the retrofit project and its implementation schedule to reduce emissions within a specific time period;

- Contains estimates of emission reductions attributable to the project, including the methodology and other technical support documentation used. EPA requires MOVES for on-road vehicles[17] and recommends NMIM for nonroad vehicles and equipment for assessing the emission reductions from retrofit projects for SIP purposes;

- Either contains federally enforceable requirements for you to implement, track, and monitor the measure; or if the measure is developed under the VMEP guidance, the state includes an enforceable commitment to monitor, assess and report the resulting emission reductions;

- If the measure is developed under the VMEP guidance, includes an enforceable commitment to remedy any SIP emission shortfall in a timely manner in the event that the measure does not achieve the estimated emission reductions; and

- Meets all other requirements for SIPs under Clean Air Act sections 110 and 172.

3.4 Are there any other types of SIPs that could include emission reductions from retrofit projects?

Yes. As indicated in Section 3.2, state air quality agencies can include retrofit projects in RFP SIPs, attainment demonstrations, and maintenance plans. However, if a state wants to encourage adoption of retrofit projects prior to developing one of these kinds of SIPs, the state could create a SIP submission that contains only retrofit projects, which would be relied upon in a future SIP. The advantage of creating such a SIP submission now is that a state could secure retrofit projects with adequate federal, state, and/or local funding for a future RFP SIP, attainment demonstration, or maintenance plan.

A state air agency that is interested in creating such a SIP submission specifically for a retrofit project or projects must consult with MPOs, the state DOT, and any other state or local transportation agencies in its development. The conformity regulation at 40 CFR 93.105 requires consultation on the development of SIPs.

---

[17] See 75 FR 9411, Official Release of the MOVES2010 Motor Vehicle Emissions Model for Emissions Inventories in SIPs and Transportation Conformity at http://www.gpo.gov/fdsys/pkg/FR-2010-03-02/html/2010-4312.htm.

Note that a SIP submission that contains only retrofit projects would still have to meet the requirements discussed in this Section. State or local agencies that are interested in this option should consult their EPA Region as well as the MPOs and others involved in the interagency consultation process.

3.5 What monitoring and record keeping should occur to document retrofit emission reductions?

3.5.1 What should the state air agency monitor and record?

Clean Air Act section 110(a)(2)(C) requires that submitted SIPs "include a program to provide for the enforcement of the measures" that the state adopts to reduce emissions. A state's decision about how a measure needs to be enforced will depend on the state's knowledge of the emission reductions achieved by the measure. Therefore, this Clean Air Act requirement for a program that provides for enforcement makes it necessary for states to monitor measures that they include in their SIPs, such as diesel retrofits.

EPA recommends that for each retrofitted vehicle, engine, or piece of equipment generating emission reductions, the state air agency or another responsible party should monitor and record the following information, where applicable, for each time period for which an emission reduction is generated:

- Actual use and operation and maintenance of the retrofitted or repowered vehicles/engines/equipment and retrofit technology
- Proper installation of retrofit technology or repowered engine at project initiation
- Proper training of vehicle operators and technicians at project initiation
- For replacements, document the permanent destruction, disabling, or rebuilding of the engine to meet current emissions standards

Monitoring and recording these data are ways to ensure that the statute is met. A state can propose other methods of monitoring and recording data in its SIP submission, and EPA would consider whether or not it would be sufficient to meet Clean Air Act requirements.

3.5.2 How long should the state air agency maintain records?

Under 28 U.S.C. 2462, the government has five years to bring an enforcement action or suit for the failure to implement a measure in a SIP. Based on this statute of limitations, all information to be monitored and recorded in accordance with this guidance for existing SIP requirements should be maintained by the state air agency or another responsible party for a period of no less than five years, or longer where appropriate.

3.6 What validation and reconciliation should occur for emission reductions in SIPs approved under the VMEP guidance?

The SIP submission for a voluntary measure should contain a description of the evaluation procedures and time frame(s) in which the evaluation of SIP reductions will take place. Once the voluntary control measure is in place, emission reductions should be evaluated by the state or local agency as required to validate the actual emission reductions. The state or local agency should submit the results of the evaluation to EPA in accordance with the schedule contained in the SIP. If the review indicates that the actual emission reductions are not consistent with the estimated emission reductions in the SIP, then the amount of emission reductions in the SIP should be adjusted appropriately or applicable remedial measures should be taken under the VMEP guidance. See EPA's VMEP guidance for further information regarding validation and reconciliation requirements for such measures.

3.7 What penalties can EPA impose for not complying with Clean Air Act requirements?

Use of this guidance does not relieve you of any obligation to comply with all otherwise applicable CAA requirements, including those pertaining to the crediting of emission reductions for your SIP, such as for your attainment demonstration or maintenance plan. Violations of CAA requirements are subject to administrative, civil, and/or criminal enforcement under Section 113 of the CAA, as well as to citizen suits under Section 304 of the CAA. The full range of penalty and injunctive relief options would be available to the federal or state government (or citizens) bringing the enforcement action.

SECTION 4: USING EMISSION REDUCTIONS IN TRANSPORTATION CONFORMITY DETERMINATIONS

4.1 What is transportation conformity?

Transportation conformity is required under Clean Air Act section 176(c) (42 U.S.C. 7506(c)) to ensure that federally supported highway and transit project activities are consistent with ("conform to") the purpose of the SIP. Conformity to the purpose of the SIP means that transportation activities will not cause new air quality violations, worsen existing violations, or delay timely attainment of the relevant NAAQS and interim milestones. EPA's transportation conformity rule (40 CFR Parts 51 and 93) establishes the criteria and procedures for determining whether transportation plans, TIPs or projects conform to the SIP. Transportation conformity applies to areas that are designated nonattainment, and those redesignated to attainment after 1990 ("maintenance areas") for transportation-related criteria pollutants. Some areas that are currently doing transportation conformity could benefit from retrofit project reductions, including ozone, $PM_{2.5}$, and PM_{10} nonattainment and maintenance areas.

In urban areas, transportation planning and conformity determinations are the responsibility of the MPO. MPOs are responsible for updating and revising the transportation plan and TIP on a periodic basis, as well as making transportation plan and TIP conformity determinations. Such a determination includes a regional emissions analysis that shows that the emissions expected from the area's planned transportation system do not exceed the motor vehicle emissions budget ("budget") set by the SIP for meeting RFP, attainment, or maintenance requirements. In cases where an area does not yet have a SIP in place, a different type of emissions test[18] is used for conformity. After an MPO's conformity determination, the U.S. DOT must also determine conformity of the transportation plan and/or TIP. The interagency consultation process is required to be used when developing transportation plans, TIPs, conformity determinations, and SIPs, and the process includes MPOs, state departments of transportation, public transit agencies, other transportation agencies, state and local air quality agencies, EPA, and DOT (40 CFR 93.105).

4.2 What kinds of retrofit projects can be used in transportation conformity determinations?

MPOs may use emission reductions in transportation plan and TIP conformity determinations that result from either:

---

[18] In areas without SIP budgets, an interim emissions test(s) must be met. These tests are the "baseline year test" and the "build/no-build test." See 40 CFR 93.119 for more on these requirements. In areas with SIP budgets, a budget test must be met, pursuant to 40 CFR 93.118.

- On-road vehicle and engine retrofit projects, or
- Nonroad vehicle, engine, and equipment retrofit projects.

On-road vehicle retrofit projects can be used in regional emissions analyses for transportation conformity determinations, since transportation conformity includes emissions and reductions from on-road sources. Requirements are discussed in Section 4.3.

Although the emissions from nonroad sources are separate from on-road vehicles in a SIP inventory, the emission reductions that result from nonroad retrofits can still be applied in transportation conformity. When appropriate and desired, the transportation conformity rule allows options for including reductions from nonroad retrofit projects, such as the retrofitting of highway construction equipment, in a transportation conformity determination. See Sections 4.4 through the end of this section for further information on options for including the emission reductions from nonroad retrofit projects in transportation conformity determinations.

EPA also notes that reductions from retrofits can be applied as a mitigation or control measure in quantitative $PM_{2.5}$ or PM_{10} hot-spot analyses for project level conformity determinations. See EPA's Transportation Conformity Guidance for Quantitative Hot-spot Analyses in $PM_{2.5}$ and PM_{10} Nonattainment and Maintenance Areas at www.epa.gov/otaq/stateresources/transconf/projectlevel-hotspot.htm#pm-hotspot. For further background on PM hot-spot analyses, the methodology in Section 2 of this guidance can be used for such analyses. The remainder of Section 4 and the Appendices focus on applying retrofit reductions in transportation plan and TIP transportation conformity determinations (and regional emissions analyses).

4.3 How can the emission reductions from on-road retrofit projects be included in transportation conformity determinations?

The transportation conformity rule describes the specific requirements for including emission reductions from on-road retrofit projects in a transportation conformity determination. If the emission reductions from the retrofit project have been accounted for in the SIP's motor vehicle emissions budget, the MPO would also include the emission reductions from the retrofit project, to the extent it is being implemented, when estimating regional emissions for a transportation conformity determination. Including the emission reductions in both the SIP's budget and in a conformity determination in this way is not "double-counting," but correctly accounting for all the control measures that are in place.[19]

To include the emission reductions from retrofit projects in a conformity analysis, the appropriate jurisdictions must be committed to the measure.[20] The appropriate level of commitment

---

[19] See 40 CFR 93.122(a) for the requirements regarding what must be included when estimating regional emissions in a conformity determination.

[20] As stated in Section 1.7 of this guidance, state and local agencies should check with EPA before promulgating any state or local regulations or programs mandating retrofit projects, due to potential issues related to Clean Air Act section 209 requirements.

varies according to the requirements outlined in 40 CFR 93.122(a) which are described as follows:

- If the retrofit project does not require a regulatory action to be implemented and it is included in the transportation plan and TIP with sufficient funding and other resources for its full implementation, it can be included in a transportation conformity determination.

- If the retrofit project requires a regulatory action to be implemented, it can be included in a conformity determination if one of the following has occurred:

    o The regulatory action for the retrofit project is already adopted by the enforcing jurisdiction (e.g., a state has adopted a rule to require such a project);

    o The retrofit project has been included in an approved SIP; or

    o There is a written commitment to implement the retrofit project in a submitted SIP with a motor vehicle emissions budget that EPA has found adequate.[21]

If the retrofit project is not included in the transportation plan and TIP or the SIP, and it does not require a regulatory action to be implemented, then it can be included in the transportation conformity determination's regional emissions analysis if the determination contains a written commitment from the appropriate entities to implement the retrofit project.

Whatever the case, any emission reductions can only be applied in a transportation conformity determination for the time period or years in which the retrofit project will be implemented. Written commitments must come from the agency with the authority to implement the retrofit project. The latest emissions model and planning assumptions must also be used when calculating emission reductions, according to 40 CFR 93.110 and 93.111.

The interagency consultation process must be utilized (as required by 40 CFR 93.105) to discuss the methods and assumptions used to quantify the reductions from the retrofit project. Section 2 of this document describes how to quantify emission reductions.

4.4 How can the emission reductions from nonroad retrofit projects be included in transportation conformity determinations?

There are two options that may be used to reflect reductions from nonroad retrofit projects in transportation conformity determinations. The two options are:

- Apply nonroad retrofit emission reductions as a "safety margin" to the SIP's motor vehicle emissions budgets; or

---

[21] 40 CFR 93.118 describes the process and criteria that EPA considers when determining whether submitted SIP budgets are appropriate for transportation conformity purposes prior to EPA's SIP approval action.

- Establish a trading mechanism in the SIP to allow emissions to be traded from one emissions sector to another.

Both of these options are allowed by the current transportation conformity rule and are completed through the SIP process with consultation among federal, state, and local air quality and transportation agencies. An area may decide to pursue one of these options if it is anticipated that emission reductions from nonroad retrofit projects may be needed to assure future transportation conformity determinations. Note that nonroad retrofit projects that are reflected in a transportation conformity determination under either option must meet the conformity rule requirements articulated in Section 4.3, in addition to other requirements described below.

4.5 How is EPA supporting the use of emission reductions from nonroad retrofit projects in transportation conformity?

EPA recognizes the importance of nonroad retrofit projects in reducing emissions and wants to support the process for state and local air and transportation agencies to implement them when desired. The remainder of Section 4 and Appendices A through C include additional information for including the emission reductions from nonroad retrofit projects in transportation conformity:

Safety margins. EPA has provided the following:
- Questions and answers about safety margins, beginning with Section 4.6;
- Step-by-step process instructions, including a flowchart, for adopting a safety margin – see Appendix A;
- An example of how a safety margin is applied – see Appendix A;
- A list of areas that have adopted safety margins in the past – see Appendix A;

Trading mechanisms. EPA has provided the following:
- Questions and answers about trading mechanisms, beginning with Section 4.11;
- Step-by-step process instructions, including a flowchart, for adopting a trading mechanism – see Appendix B;
- An example of how a trading mechanism would work – see Appendix B;
- A model trading rule that interested states could adopt through the SIP – see Appendix C.

4.6 What is a safety margin?

Section 93.101 of the transportation conformity rule defines a "safety margin" as

"the amount by which the total projected emissions from all sources of a given pollutant are less than the total emissions that would satisfy the applicable requirement for reasonable further progress, attainment, or maintenance."

That is, if an area has a safety margin, it has more emission reductions than necessary to meet the

Clean Air Act goal of RFP, attainment, or maintenance. It can meet the goal with room to spare – a margin of safety. Safety margins are calculated for a specific year in the SIP for which a budget is established (e.g., the last year of a maintenance plan).

The conformity rule allows a safety margin to be allocated to the transportation sector, if the SIP explicitly states so (see 40 CFR 93.124(a)). The SIP must state that a specified portion or the entire safety margin is available to the MPO and DOT for conformity purposes. Therefore, the nonroad retrofit project that creates the safety margin allows the motor vehicle emissions budget to increase.

4.7   When can a safety margin be established in a SIP?

An area may include a safety margin from the point at which they develop their SIP submission for RFP, attainment, or maintenance, or it may be developed later as a stand-alone SIP submission. For example, if an area has more emission reductions from the control measures contained in its attainment demonstration than needed to achieve the applicable NAAQS, it could establish a safety margin in that attainment demonstration.

An area could also develop a safety margin after the SIP has been established. For example, if an area's attainment demonstration contains control measures that are sufficient to attain, then any other measures that reduce emissions adopted later, such as nonroad retrofit projects, would potentially create a safety margin. Alternatively, an area that already has a safety margin in its SIP could add to it by adopting a new nonroad retrofit project.

Whatever the case, any SIP that incorporates a safety margin must meet all applicable SIP requirements, including being based on the most recent emissions estimates available at the time the SIP safety margin is developed. For example, an existing safety margin in an ozone maintenance plan could be increased to account for new retrofit projects in a given year, assuming the existing safety margin is still applicable. Under this example, EPA's approval would be required before the increased SIP safety margin could apply for transportation conformity purposes, since revisions to existing approved SIPs cannot apply until EPA approves them (40 CFR 93.118(e)(1)).

4.8   What are the benefits of establishing a safety margin at the time that an RFP SIP, attainment demonstration, or maintenance plan is initially submitted?

While a safety margin can be developed after a SIP is established, it makes sense for areas that are currently developing a SIP to include a safety margin in that SIP, if possible, instead of adding it later. A safety margin that is included in an initial control strategy SIP (e.g., an RFP SIP or attainment demonstration) or maintenance plan could be used for conformity on the effective date of EPA's adequacy finding. But if a safety margin is developed later, the state would need to revise the SIP. If EPA has already approved the initial SIP, the revised SIP that includes the safety margin could not be used in conformity until it is approved, as noted above. Including a safety margin in a SIP submission as it is developed saves time and effort throughout the process.

4.9 Why should an area start developing a safety margin now?

Areas that are currently developing attainment demonstrations or maintenance plans should consider whether they want to adopt programs to retrofit nonroad vehicles, engines, and equipment and include them in these SIP submissions, because doing so now can save time and may allow their use in conformity sooner. The conformity rule requires that consultation occur during the development of a SIP (40 CFR 93.105(a)(1)). State air agencies should consult with the relevant MPOs, DOT, and other agencies as appropriate in order to determine whether a safety margin might be needed for conformity.

Several nonattainment areas have already established safety margins through the SIP process to assist in making future transportation conformity determinations. See Appendix A for further information.

4.10 What are the features of a safety margin?

In order for EPA to approve the allocation of a safety margin to a motor vehicle emissions budget(s), the following SIP and transportation conformity requirements would have to be met:

- The entire SIP must continue to demonstrate its Clean Air Act purpose, pursuant to the statute and 40 CFR 93.124(a). Before the emissions level of a motor vehicle emissions budget is increased, the state air agency would need to determine that there is a safety margin, including any surplus emission reductions from any nonroad retrofit projects. The agency would also need to ensure that emissions inventories of on-road, nonroad and other sources are consistent with the SIP's demonstration.

- The calculation for the nonroad retrofit projects and any safety margin would be based on the latest information and models available at the time the SIP is developed. The method for calculating a safety margin may vary depending upon whether new air quality modeling is performed or if a less rigorous demonstration is adequate (e.g., maintenance areas that are establishing a safety margin based on staying below the emissions level(s) for a previous year of clean monitoring data for the maintenance demonstration).

- The SIP must clearly allocate the safety margin to the motor vehicle emissions budget(s) for use in transportation conformity determinations, pursuant to 40 CFR 93.124(a);

- Retrofit projects reflected in a safety margin must be assured, permanent, and enforceable, have adequate funding and resource commitments, and be on schedule; and

- The safety margin must meet Clean Air Act section 110(l) and any other applicable SIP statutory and regulatory requirements.

See Section 3 for more information about including a retrofit project in a SIP. See Appendix A for more detailed information on establishing and implementing a safety margin.

EPA notes that the allocation of a safety margin to the on-road transportation sector may limit that area's ability to allow future growth in emissions from other source sectors (e.g., stationary sources). State and local transportation and air quality agencies and other affected parties should always consult on whether a safety margin is appropriate for transportation conformity in a given area.

4.11 What is a trading mechanism?

A trading mechanism is a process established through the SIP that allows emission reductions achieved in another source sector – such as the nonroad sector – to be used for demonstrating transportation conformity. Before emission reductions can be traded from one sector to another, the SIP must include a trading mechanism to allow the trading to occur.

This option is supported by 40 CFR 93.124(b) as well as preamble language. Section 93.124(b) states:

> "A conformity demonstration shall not trade emissions among budgets which the applicable implementation plan (or implementation plan submission) allocates for different pollutants or precursors, or among budgets allocated to motor vehicles and other sources, unless the implementation plan establishes appropriate mechanisms for such trades."

In the preamble of the November 24, 1993 transportation conformity rule, EPA stated that "[t]he state may choose to revise its SIP emissions budgets in order to reallocate emissions among sources or among pollutants and precursors. For example, if the SIP is revised to provide for greater control of stationary source emissions, the State may choose to increase the motor vehicle emissions budget to allow corresponding growth in motor vehicle emissions (provided the resulting total emissions are still adequate to provide for attainment/maintenance of the NAAQS..." (58 FR 62196). EPA believes that this preamble and 40 CFR 93.124(b) clearly allow trading mechanisms to be established to ensure future transportation conformity determinations, when desired.

EPA's Economic Incentive Program (EIP) guidance also discusses trading among different sources.[22] The primary concerns articulated in that guidance are: 1) to ensure that emission reductions are only used in transportation conformity if they are truly reductions to overall air

---

[22] EPA's January 2001 guidance entitled, "Improving Air Quality with Economic Incentive Programs," provides additional information on developing and implementing economic incentive based control strategies. This guidance is available at: www.epa.gov/ttn/oarpg/t1/memoranda/eipfin.pdf. See Appendix 16.10 of the 2001 guidance for further information regarding the transportation conformity requirements that need to be met by an EIP trading program.

quality (i.e., are not offset by emissions increases in other emission sources); and 2) to ensure that double-counting does not occur among source categories in the SIP or in transportation conformity. EPA must approve a SIP that creates a trading mechanism before emission reductions can be traded for transportation conformity determinations.

4.12 When can a trading mechanism be established in a SIP?

A trading mechanism can be established by a state and approved by EPA into the SIP for a nonattainment or maintenance area at any time.

- A trading mechanism that would allow trading of emission reductions from nonroad retrofit projects for transportation conformity purposes could be submitted by a state and approved by EPA prior to the submission of an area's RFP SIP, attainment demonstration, or maintenance plan.

- Alternatively, a state could decide to submit a trading mechanism as part of an RFP SIP, attainment demonstration, or maintenance plan for a given area.

4.13 Why develop a trading mechanism as a stand-alone SIP revision?

The advantage of developing a trading mechanism as a stand-alone SIP instead of including it in a required ozone or a $PM_{2.5}$ SIP is that doing so allows trading to happen sooner than if the trading mechanism was combined with a larger SIP such as an RFP plan, an attainment demonstration, or a maintenance plan. Where developed as a stand-alone SIP, a trading mechanism allows emission reductions from a nonroad retrofit project to be applied in conformity during the time period before one of these larger SIPs is submitted and found adequate. For example, assuming it has a SIP trading mechanism in place, if an ozone area could not pass a NOx budget contained in its ozone attainment demonstration, the area could program funds for a nonroad retrofit project and apply the emission reductions in a conformity determination. A stand-alone SIP based on the model trading rule provided in this guidance document can be developed and approved by EPA through parallel processing[23] the SIP.

4.14 What would a SIP with a trading mechanism look like?

EPA developed a model trading rule to allow states to use nonroad emission reductions in

---

[23] The term "parallel processing" means that EPA, at the request of the state, begins to process a SIP submission (either to approve it or find its budgets adequate) before the state has finalized it. EPA's process and the state's process are occurring "in parallel." Specifically, EPA would begin to review the SIP trading mechanism while the state's public comment period is running, but must wait until the state submits its final SIP trading mechanism before taking a final SIP action. See Appendices A and B for more detail on how parallel processing works.

transportation conformity. States that are interested in establishing a trading mechanism merely need to copy the model rule and fill in the blanks as appropriate (e.g., the state would insert the appropriate names of state air quality agencies and MPOs), and can adopt it into their SIPs. The model rule is found in Appendix C of this guidance document. EPA developed this model rule in consultation with DOT and state and local air quality and transportation organizations. States that adopt the model rule can expect a streamlined SIP approval because the model rule meets Clean Air Act requirements. The model rule is just one way to write a regulation for a trading mechanism for conformity. A state could deviate from this model rule, but the more the state deviates from the model rule language, the more time it may take for EPA's final SIP action.

4.15 What are the features of a trading mechanism?

In order for EPA to approve a trading mechanism, the following SIP and transportation conformity requirements would have to be met:

- The entire SIP must continue to demonstrate its Clean Air Act purpose when trades are allowed, pursuant to the statute and 40 CFR 93.124(b). When the trading mechanism is developed, the state air agency would need to ensure that trades can occur, while ensuring that emissions inventories from all sources continue to be consistent with the SIP's demonstration.

- A state regulation must be developed that describes in adequate detail the scope and process for making trades so that the trading mechanism can be implemented as intended. Such a regulation would also provide that all trades become enforceable through the trading mechanism approved into the SIP in the event that reductions are not otherwise enforceable.

- Trades must be based on nonroad retrofit projects that result in a net emission reduction, which must consider the emissions produced by the vehicles, engines, and equipment and the impact of any retrofit technology. Individual trades cannot exceed the net emission reductions of a retrofit project.

- The trading mechanism would ensure that individual trades must be made only when surplus emission reductions exist. Trading cannot result in "double-counting" of reductions already accounted for in the SIP and conformity determinations, .

- Reductions can be taken only for years of the regional emissions analysis where reductions are achieved.

- The trading mechanism must meet any other applicable SIP and conformity requirements (for example, control programs that produce emission reductions relied on in a conformity determination must be assured and enforceable, have adequate funding and resource commitments, and be on schedule – see 40 CFR 93.122(a)).[24]

---

[24] EPA notes that SIPs that establish a trading mechanism are required to meet Clean Air Act section 110(l) requirements, similar to all other SIPs for other purposes. However, by definition, the trading mechanisms described in

See Appendix B for more information on implementing a trading mechanism and Appendix C for a model trading rule that could be incorporated into the SIP.

4.16 Does the public have the opportunity to comment on trading?

Yes, the public does have the opportunity to comment at two distinct points in the process:

- First, the public can comment on the trading mechanism itself at the point when it is incorporated into the SIP. Any time a state prepares a SIP, it must give the public an opportunity to comment on it before the final submission is sent to EPA. In a situation where a state adopts a regulation separately from an RFP or attainment SIP or a maintenance plan, the public will also have the opportunity to comment on the state's regulation. In addition, when EPA proposes to approve the SIP in the Federal Register, the public again has the opportunity to comment to EPA. The approval process, including these public comment periods, is covered in further detail in Appendix B.

- Second, the public has the opportunity to comment on individual trades that employ the SIP trading mechanism during the public comment period on a transportation plan or TIP conformity determination. The conformity regulation at 40 CFR 93.105(e) requires that the public be given access to technical and policy information considered in a conformity determination at the beginning of the public comment period and prior to taking formal action on a transportation plan/TIP conformity determination. If an MPO relies on nonroad retrofit emission reductions that are the result of a trade, information about that trade would belong in such technical and policy information that is made available for public review. Under the model trading rule in Appendix C of this document, the MPO would be required to document the description of the retrofit project; its enforceability, funding sources, and implementation schedule; and details regarding emission reduction calculations. In addition, the model rule would require that the state air agency document its concurrence (e.g., through a letter), which the MPO could then reference in its transportation plan/TIP conformity determination. If a state adopts any trading mechanism, all of the information described above would be made available to the public as part of any transportation plan/TIP conformity determination.[25]

4.17 Are there any examples of transportation conformity-related trading mechanisms that were approved into the SIP?

---

this guidance document should meet section 110(l) requirements since emissions increases would not occur as a result of trades for transportation conformity purposes.

[25] Note that if a state decides to substantially alter the model rule or create a different one, the trading mechanism would still need to provide the public with access to information via the conformity determination to meet 40 CFR 93.105(e). EPA would examine any trading mechanism with regard to this point during the SIP process.

Yes, a couple of areas adopted another type of trading mechanism for conformity purposes. The provision of the transportation conformity rule that allows trading among source sectors also allows trading among motor vehicle emissions budgets of pollutants and precursors. Both the San Joaquin Valley and Salt Lake City adopted trading mechanisms that established trading between budgets for directly-emitted PM_{10} and NOx.[26]

Though these trading mechanisms applied to the trading of on-road mobile precursors and pollutants rather than among source sectors, they have provided EPA with valuable experience in developing a trading mechanism through the SIP process. EPA has applied this experience to the development of the model trading rule found in Appendix C. In addition, Salt Lake City's trading mechanism was needed as soon as it could be approved, so EPA and the State of Utah worked together to expedite its approval into the SIP.

4.18 Are there any other ways to establish a trading mechanism for a nonroad retrofit project?

Yes. Sections 4.11 through 4.17 and Appendices B and C address establishing a trading mechanism that, once adopted into a SIP, would allow emission reductions from any nonroad retrofit project to be traded to the transportation sector. However, another possibility is for a state to adopt a SIP that allows trading of emission reductions from only a specific project (or projects). This type of SIP would look similar to the type of SIP submission described in Section 3.4 in that it would contain the description of, and emission reductions from, a specific project (or projects). It would also explicitly state the portion of the retrofit emission reductions that are available to be used in transportation conformity determinations. The advantage of creating such a SIP submission now is that a state could secure specific retrofit projects with adequate federal, state, local, or private funding for a future conformity determination. Where a SIP specifically states that a certain number of tons of retrofit reductions are to be preserved for use in future conformity determinations, such reductions would not be available for use in any other SIP demonstrations such as RFP or attainment demonstrations.

State or local agencies that are interested in this option should consult with their EPA Region as well as the MPOs and others involved in the interagency consultation process.

4.19 How can I get additional technical assistance in using the emission reductions from a nonroad retrofit project in transportation conformity?

---

[26] These trading mechanisms differed according to the atmospheric conditions in each area. An interpollutant trading mechanism must include a scientific rationale, as well as comply with Clean Air Act section 110(l) by showing that the trading mechanism will not interfere with any applicable Clean Air Act requirement. See 67 FR 44065 for EPA's approval of Salt Lake City's trading mechanism; 69 FR 30006 for EPA's approval of San Joaquin Valley's PM_{10} SIP that includes the trading mechanism.

State and local agencies are strongly encouraged to consult early with their respective EPA Region if they are considering applying a safety margin or developing a trading mechanism in the SIP for transportation conformity purposes. EPA is available to provide such technical assistance so that SIP submissions can be processed efficiently. See Section 1.9 for more information regarding EPA contact information.

SECTION 5: USING EMISSION REDUCTIONS IN GENERAL CONFORMITY DETERMINATIONS

5.1 How can the estimated emission reductions be used for general conformity determinations?

General conformity applies in nonattainment and maintenance areas to emissions from all federal actions or activities not covered by the transportation conformity program, such as military base expansions or approval of airport expansion projects. The general conformity regulation (40 CFR 51.851 and 93.150-165) prohibits federal agencies from taking, or supporting in any way, actions in areas that are designated nonattainment or maintenance for any criteria pollutant without demonstrating that the action will not 1) cause or contribute to any new violation of any NAAQS in any area; 2) increase the frequency or severity of any existing violation of any NAAQS in any area; or 3) delay timely attainment of any NAAQS or any required interim emission reductions or other milestones in any area. The general conformity regulation applies to the total foreseeable direct and indirect emissions increases from the action or activity including both construction and operational emissions. Indirect emissions can include those from vehicles servicing a federal facility or activities supported by federal funds.

To reduce the regulatory burden on insignificant actions, the regulation establishes a number of exemptions for categories of actions or activities known to have insignificant emissions increases or whose emissions fall below certain *de minimis* levels. The general conformity rule establishes *de minimis* emission levels based on the severity of the nonattainment problem. If the net increase in total direct and indirect emissions from a federal action is below the *de minimis* levels, the federal agency does not have to make a conformity determination for the action.

If the net increase in emissions is above the *de minimis* levels and the project is not otherwise exempt, the federal agency must determine that the action or activity will conform to the SIP. Because the general conformity regulation applies to a wide variety of actions or activities, the rule provides a number of methods to demonstrate conformity. Two methods, mitigation and offsetting of emissions increases, require emission reductions from sources which may, or may not be directly connected to the federal action or activity.

5.2 How can retrofit projects be used to meet the general conformity requirements?

There are some fleets of vehicles that are not considered in transportation conformity determinations, and instead considered in general conformity determinations. Some examples include a fleet of trucks at a military base, or vehicles that operate only within the boundary of a commercial airport. Since a retrofit project for such a fleet would not be increasing emissions, the implementation of a retrofit project would not be subject to general conformity requirements. Additionally, in many cases, a federal action would not be needed to implement such a project.

However, emission reductions generated by diesel retrofit projects could be used in a number of ways in general conformity determinations, such as:

- Diesel engine retrofits could be used to mitigate or offset emissions increases caused by a federal action. For example, the retrofitting of diesel airplane tugs at a military air station could be used to mitigate a portion of the emissions increases associated with an expansion project at the military base or the retrofitting of package delivery vehicles could be used to offset the emissions increases caused by a base expansion.

- The Vision 100-Century of Aviation Reauthorization Act of 2003 (P.L. 108-176), directed the FAA to establish a national program to reduce airport ground emissions at commercial service airports located in air quality nonattainment and maintenance areas. The Voluntary Airport Low Emissions (VALE) program allows airport sponsors to use certain funds to finance low emission vehicles, refueling and recharging stations, gate electrification, and other airport air quality improvements (including retrofit projects). The emission reductions generated by these measures are kept by the airport sponsor and may only be used for current or future general conformity determinations. To be used in a general conformity determination, emission reductions must be voluntary and cannot otherwise be used to meet other applicable air quality attainment and maintenance requirements. Emission reductions must also be "permanent" in that they continue to occur at the estimated level throughout the lifetime of the vehicles and infrastructure.

    A federal or federally permitted facility which is subject to the general conformity regulation, such as a military base or a commercial airport, could institute an agreement with a state to operate the facility within a facility-wide emissions budget. Emission reductions from a fleet retrofitting project could ensure emissions increases generated by future actions would not exceed the facility-wide budget and thus would conform to the SIP. A demonstration that emissions from a new action do not exceed the facility-wide emissions budget could be used in a general conformity determination.

5.3 *Is a SIP revision required if a source with a facility-wide emissions budget wants to implement a retrofit project at the facility?*

No, a SIP revision would not be required, if a source with a facility-wide emissions budget established for general conformity purposes decided to retrofit vehicles or equipment used at the facility. The facility's emissions budget would have included emissions of all relevant pollutants and/or precursors from all sources on the facility including the pre-retrofit emissions from the vehicles and/or equipment that are included in the retrofit project. Therefore, the emission reductions from a retrofit project can be used to show that the facility remains within its budget without a SIP revision.

5.4 What requirements would potentially limit the use of retrofit projects in the general conformity program?

Emission reductions used as part of the (1) project design, (2) mitigation measures, (3) offset, or (4) in future conformity determinations, must be surplus, permanent, quantifiable and enforceable as described in the General Conformity rule and Section 3 of this document. In addition, the emission reductions must be reviewed by EPA, state, tribes and local air quality agencies and the public as part of the review of the general conformity determination and meet the following criteria:

- The retrofit project must be identified and the process for implementation and enforcement must be explicitly described.

- Prior to determining conformity, the federal agency making the determination must obtain written commitments from the appropriate entities (e.g., fleet operator, state or city official, private company official or MPO) to implement the retrofit project as a mitigation measure.

- The reductions from the retrofit project must be contemporaneous with the project emissions increases, specifically, the reductions must occur in the same calendar year as the increases.

- The implementing entity/official responsible for implementing the retrofit project and any persons or agencies voluntarily committing to mitigation measures must comply with the obligations of such commitments.

- If the federal action involves licensing, permitting or approving an action of another governmental or private entity, the federal agency must condition its approval action on the other entity meeting all mitigation commitments.

- If the retrofit project is modified resulting in an increase in emissions, the new mitigation measures must continue to support the initial general conformity determination and must undergo public review.

5.5 Why are SIPs required for use of nonroad retrofits in transportation conformity but not required for general conformity?

With regard to nonroad retrofit projects, a SIP establishing either a trading rule or safety margin is required in order for emission reductions from such projects to be used in transportation conformity determinations because nonroad emissions are not included in an area's on-road motor vehicle emissions budget (40 CFR 93.124(a) and (b)). The SIP provides the authority to use nonroad emission reductions as part of a demonstration that an area is meeting its on-road emissions budgets. In contrast, a facility-wide emissions budget established for general conformity purposes would include emissions from all sources at the facility including the pre-retrofit emissions from the vehicles and/or equipment that are included the retrofit project and the facility budget is within the SIP's allowable emission budget. Therefore, the emission reductions from the retrofit project can be used to show that the facility remains within its budget without a SIP revision.

APPENDIX A:

ESTABLISHING AND USING A SAFETY MARGIN TO ALLOW EMISSION REDUCTIONS FROM NONROAD RETROFITS TO BE USED IN TRANSPORTATION CONFORMITY DETERMINATIONS

This appendix provides detail about how a state could apply the emission reductions from a nonroad retrofit project or projects in a conformity determination via a safety margin.

A.1 What is an example of establishing and using a safety margin?

Suppose a 1997 $PM_{2.5}$ nonattainment area develops a draft maintenance plan that shows the area could emit 400 tons per day of NOx from all sources, and still maintain the NAAQS in the last year of the maintenance plan, 2025. The area's total NOx SIP inventory for 2025 is:

Nonroad:	100 tons per day
On-road:	100 tons per day
Stationary sources:	100 tons per day
Area sources:	100 tons per day
Initial total maintenance SIP inventory:	400 tons per day

If the area submitted this draft SIP without a safety margin, this SIP would have a 2025 NOx motor vehicle emissions budget of 100 tons per day, which is the on-road portion of the 2025 NOx inventory.

However, the MPO, in consultation with others, decides to spend some of its CMAQ and other funding to retrofit nonroad transportation construction equipment. The MPO consults with the state air agency, who agrees that the retrofits could be used to create a safety margin, since the area can demonstrate maintenance without these additional reductions. The MPO decides to retrofit a number of bulldozers, front loaders, and backhoes and estimates that the total emission reductions from these retrofits will be 5 tons per day in 2025. The state air agency reviews these calculations and concurs that these reductions are surplus. Therefore, after consultation, it adjusts the NOx inventory in its draft maintenance plan for 2025 as follows:

Nonroad:	95 tons per day (i.e., 100 minus 5 tons for new retrofit)
On-road:	105 tons per day (i.e., 100 plus 5 tons for new retrofit)
Stationary sources:	100 tons per day
Area sources:	100 tons per day
Final total maintenance SIP inventory:	400 tons per day

The state submits a final SIP that includes a 5 ton per day safety margin and specifically states in the SIP that it is allocated to the motor vehicle emissions budget. Therefore, the final budget is 105 tons per day:

On-road:	100 tons per day
Safety margin:	5 tons per day
Total available for transportation:	105 tons per day (the final budget)

Once the SIP has been submitted and found adequate by EPA, the MPO uses the new 2025 NOx budget of 105 tons per day. This new budget would apply to all conformity determinations for all analysis years 2025 and later, for as long as this motor vehicle emission budget is in place without other future budgets. If the MPO later wanted to fund additional nonroad retrofit projects and use the additional reductions for conformity determinations, they would either have to work with the state to revise the safety margin in the SIP or implement a SIP trading mechanism (as discussed in Section 4 and Appendix B).

A.2   *What is the most expeditious process for establishing a safety margin that results from a nonroad retrofit project?*

An area that wants to include a safety margin in its SIP from a retrofit project or projects would follow the following steps. (See Figure 1 at the end of this appendix for a flowchart that depicts these steps.) These steps assume that EPA has not yet approved a SIP submission that establishes budgets that the safety margin amends, that is, the safety margin is included in the initial RFP SIP, attainment demonstration, or maintenance plan that the state submits. Therefore, the budgets that include the safety margin can be used as soon as EPA finds them adequate. However, if the area has an approved SIP with a motor vehicle emissions budget and the state submits a SIP with a safety margin that would change the budget, the revised budget could not be used until EPA approves the SIP.

- **Agencies consult and decide on the project.** The process would begin with the MPO and state air agency discussing the ability to create a safety margin by retrofitting nonroad vehicles, engines, and equipment and the scope of the retrofit project(s) that could contribute to the safety margin. There may be some areas where a safety margin cannot be created; in these cases nonroad retrofit projects may be even more important for the area's SIP. Assuming that the nonroad retrofit project is funded at least in part with CMAQ dollars, the MPO (or state DOT) would need to determine the timing of the project, number of vehicles, engines, or equipment to be retrofitted, and the retrofit technology that would be applied because the MPO (or state DOT) determines how CMAQ money is spent. The state air agency also must be involved because that agency prepares the SIP.

- **Agencies quantify the project.** The state air agency, in consultation with the MPO,[27] would

---

[27] Note that either the state air agency or the MPO could take the lead role in this step.

quantify emission reductions from the retrofit project in the year(s) for which the budgets are established. (See Section 2 of this guidance for further direction in how a retrofit project can be quantified.)

- **State air agency assures SIP criteria are met.** The state air agency would determine and document that the emission reductions meet all of the criteria necessary for a SIP to include emission reductions from a project, specifically, that emission reductions are surplus, quantifiable, enforceable, permanent, and adequately supported. (See Section 3 of this guidance for further discussion of these criteria.)

- **State air agency prepares the safety margin section of the SIP.** The state air agency would prepare an initial draft of the safety margin section of the SIP; share it with the other parties in the interagency consultation process, including the MPO, DOT, and EPA; and revise it as necessary once it receives comments from the other agencies. This step could take anywhere from one to three months, depending on the nature of the comments that arise through interagency consultation. However, if the nonroad retrofit project is well thought-out and accurately quantified in the initial draft, it is likely that fewer issues would have to be resolved in the consultation process. We believe that the state air agency and the MPO could complete the first four steps in one to three months.

- **State requests that EPA parallel process the SIP.** The state air agency would send the EPA region the proposed SIP and in its transmittal letter, request that EPA use parallel processing to approve the revision and find the budgets adequate.

- **State publishes proposal.** The state must publish its proposal in accordance with its own state's procedures, which would begin a public comment period. In most cases, a state would give the public one month to comment, but depending on the circumstances, the public comment period could be shortened to 15 days.[28]

- **EPA begins the adequacy process.** Once EPA receives an initial SIP submission for parallel processing, EPA must review it for adequacy, regardless of whether the safety margin is part of a larger SIP demonstration or whether it is a stand-alone submission. EPA uses the criteria in 40 CFR 93.118(e) to judge whether a SIP submission is adequate, and follows the process described in 40 CFR 93.118(f). EPA begins the adequacy process by notifying the public that we have received a SIP submission and beginning a 30-day public comment period.

- **State revises its proposal.** The state would revise its proposal as necessary, based on the comments it receives during the public comment period. This step may take the state about a month. At this point, the state air agency finalizes the SIP and submits it to EPA for approval and completion of the adequacy process.

- **EPA completes the adequacy process.** After the comment period, EPA must inform the

---

[28]Requirements for public hearings on SIPs can be found at 40 CFR 51.102. Section 51.102(g) allows a state to shorten its public comment period if it submits a written application to EPA in advance.

state of its decision on adequacy and respond to any comments it has received. EPA also must publish a notice in the Federal Register. Usually, EPA completes the adequacy process within three months, but can expedite it if necessary to a month and a half.

- **EPA adequacy finding effective.** The adequacy finding is effective 15 days from the date the Federal Register notice is published. At this point, the MPO can use the motor vehicle emissions budgets in the SIP submission for conformity.

A.3 Have any areas included safety margins in their SIPs?

Yes. Many areas have included safety margins in their SIPs that have been used in conformity. Below we have listed areas with a safety margin as of the date of this guidance document, as well as the pollutant and the type of SIP that included the safety margin. These safety margins have resulted from a variety of control measures, rather than nonroad retrofit projects. However, the list demonstrates that safety margins have been done with some frequency in the past, and that many states have experience with including safety margins in their SIPs.

Area	Pollutant[29]	Type of SIP
Syracuse, NY	CO	maintenance plan
TN portion of the Clarksville-Hopkinsville	ozone	maintenance plan
Greene and Jackson Counties, IN	ozone	maintenance plan
Evansville, IN	ozone	maintenance plan
Terre Haute, IN	ozone	maintenance plan
Muncie, IN	ozone	maintenance plan
Colorado Springs, CO	CO	maintenance plan
Denver, CO	CO, PM_{10}	maintenance plans
Fort Collins, CO	CO	maintenance plan
Greeley, CO	CO	maintenance plan
Longmont, CO	CO	maintenance plan
Ogden, UT	CO	maintenance plan
Provo, UT	CO	maintenance plan
Salt Lake City, UT	CO	maintenance plan
San Diego, CA	ozone, CO	maintenance plans
Bakersfield, CA	CO	maintenance plan
Chico, CA	CO	maintenance plan
Fresno, CA	CO	maintenance plan
Lake Tahoe North Shore, CA	CO	maintenance plan
Lake Tahoe South Shore, CA	CO	maintenance plan
Modesto, CA	CO	maintenance plan
Sacramento, CA	CO	maintenance plan

---

[29] EPA notes that safety margins were also included in maintenance plans for the 1-hour ozone NAAQS by 14 areas.

San Francisco-Oakland-San Jose, CA   CO   maintenance plan
Stockton, CA   CO   maintenance plan

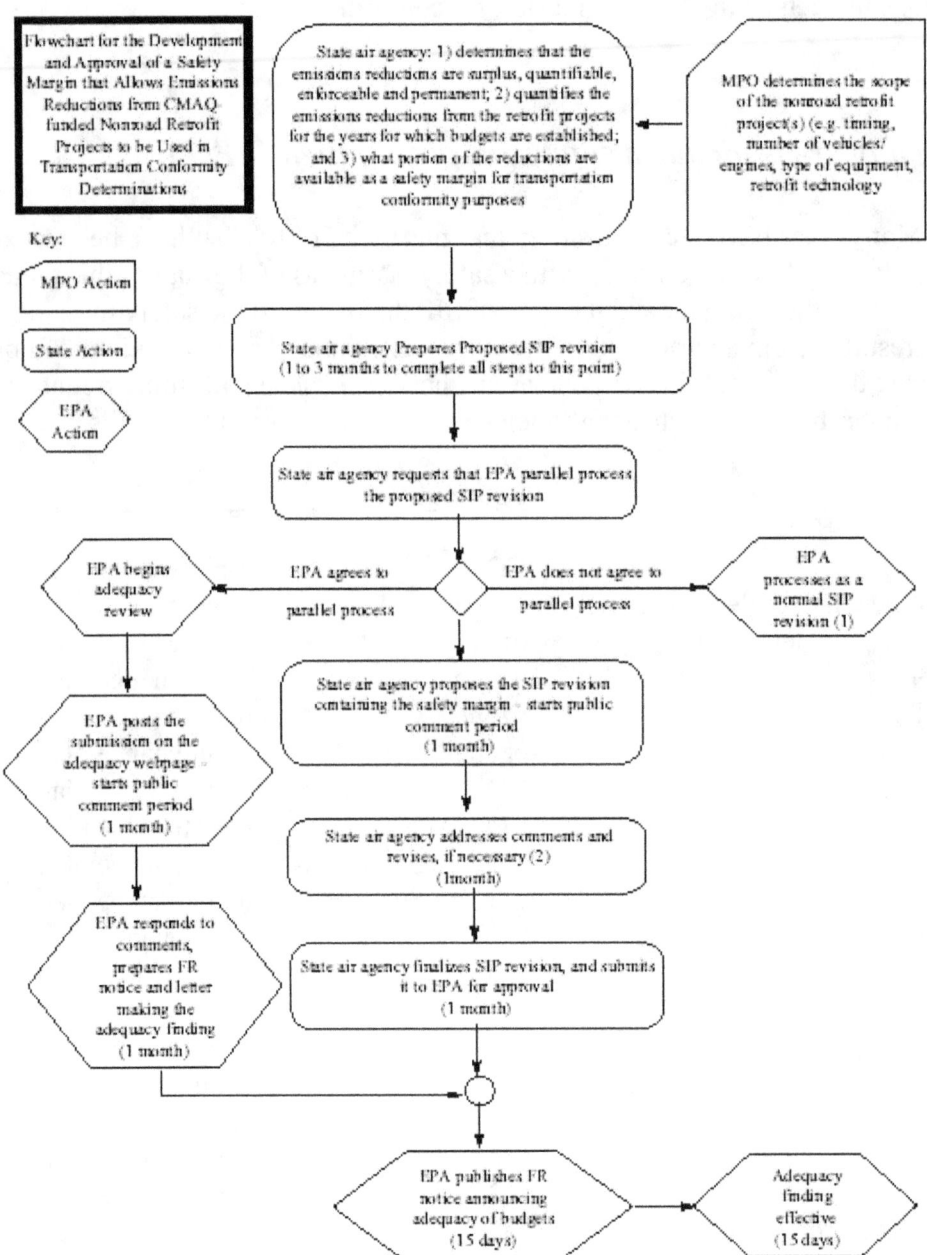

(1) EPA would not begin the adequacy process until after state submits final SIP revision for approval. Adds 3 months to the process.

(2) Assumes budgets are not affected by any revisions.

APPENDIX B:

Establishing and Using a Trading Mechanism to Allow Emission Reductions from Nonroad Retrofits to Be Used in Transportation Conformity Determinations

This appendix provides detail about how a state could apply the emission reductions from a nonroad retrofit project or projects in a conformity determination via a trading mechanism.

B.1 What is an example of establishing and using a trading mechanism?

Suppose a 2008 ozone nonattainment area develops a draft attainment demonstration that shows the area could emit from all sources 400 tons of NOx per day, and still meet the NAAQS in 2015. The area's total NOx SIP inventory for 2015 is as follows:

Nonroad:	100 tons per day
On-road:	100 tons per day
Stationary sources:	100 tons per day
Area sources:	100 tons per day
Total final attainment SIP NOx inventory:	400 tons per day

This SIP would have a 2015 NOx motor vehicle emissions budget of 100 tons per day, which is the on-road portion of the NOx inventory. The area also includes a trading mechanism in its SIP, and EPA has approved it.

During the development of the SIP, the MPO implements a nonroad retrofit project that will reduce NOx by 5 tons per day in 2015 and 2 tons per day in 2020. The state air agency and MPO concur that these reductions will not be applied to the nonroad inventory used in the SIP, in order to use it in a future transportation plan/TIP conformity determination. The state air agency concurs through a letter to the MPO that the reductions in 2015 and 2020 are surplus.

In this case, the on-road motor vehicle emission budget does not change. In subsequent conformity determinations for 2015, the MPO could offset as much as 5 additional tons of on-road emissions using the reductions from the nonroad retrofit project in 2015 and as much as 2 tons per in 2020, based on the latest information available for the retrofit project's reductions when transportation plan/TIP conformity is determined. Retrofit reductions from a trade would be estimated based on the latest assumptions and emissions model available at the time of a transportation plan/TIP determination (40 CFR 93.110 and 93.111).

B.2 Does a trading mechanism have to be included in a SIP?

Yes, a state must adopt the trading mechanism (such as the model rule found in Appendix C) into its SIP. A state may choose to develop this type of trading mechanism either as a stand-alone SIP or as part of a larger SIP submission.

EPA believes that the most expeditious process for developing and approving a trading mechanism is for a state to use the model trading rule provided in this guidance document in Appendix C to develop the mechanism as a stand-alone SIP and for EPA to parallel process its approval. As discussed below, it is possible that a trading mechanism based on the model rule may be developed by a state and approved by EPA in as little as eight months if it is submitted as a stand-alone SIP and if EPA parallel processes its approval. The model rule should be used as the basis for a state's trading rule whether the state submits its rule as a stand-alone SIP or includes it in an RFP SIP, attainment demonstration, or maintenance plan to expedite the development, review and approval of the trading mechanism.

B.3 What is the most expeditious process for establishing a trading mechanism in a SIP, and how long will it take?

To minimize the time needed for the state air agency to develop and EPA to review a SIP addressing a trading rule, EPA developed a model rule that is discussed in detail in Appendix C of this document. States that want to minimize the time necessary to adopt a trading mechanism into their SIP should use EPA's model rule. If a state decides to draft its own trading rule, it will likely take more time to complete the steps outlined below.

The following paragraphs and the flowchart at the end of this appendix (Figure 2) describe a process that can be used to expedite the approval of a trading rule into a SIP. If a state adopts the model trading rule and EPA approves the rule in parallel with the state's adoption of the rule, we believe that many states could put a trading rule in place for use in transportation conformity determinations in as little as eight months.[30]

- **State prepares proposed SIP.** The first step in developing a SIP for a trading rule to allow emission reductions from nonroad retrofit projects to be used in transportation conformity determinations is for the state air agency to prepare the proposed SIP. To expedite this step, the air agency should start with the model trading rule and modify it only to the extent necessary. The air agency would need to fill in the information pertinent to the nonattainment and maintenance areas in its state where the rule would apply. It is also likely that the state air agency would have to format the rule so that it conforms to the state's requirements for

---

[30]Some states may be able to further reduce the time needed if state law allows them to establish a SIP trading mechanism through a memorandum of agreement (MOA) or memorandum of understanding (MOU). EPA believes that use of an MOA or MOU could save time because an MOA or MOU may require less time for state reviews than a regulation. However, EPA notes that this route may not be available to all states primarily because individual state law may not allow MOAs or MOUs to be binding on future administrations.

regulatory text. The state would also develop language to be included in the preamble of its notice of proposed rulemaking. In drafting its preamble language the state may find it useful to incorporate as much relevant language as possible from the explanatory text in Appendix C that accompanies the model rule.

Once the air agency has drafted its regulation it would share the draft with the interagency consultation partners as required by 40 CFR 93.105(b)(1). The interagency consultation partners include representatives of the MPO(s), state and local air quality and transportation agencies, and the local or regional offices of EPA, FHWA, and FTA. After the interagency consultation partners have reviewed and commented on the draft, the state air agency would make any needed revisions. We believe that a state air agency could accomplish these tasks in about two months, or perhaps even less, if it closely follows the model rule in Appendix C. However, if a state air agency decided to make significant changes or to ignore the model rule, these initial steps could take up to eight months.

- **State submits proposed SIP to EPA and requests parallel processing.** At this point the air agency would transmit the SIP to the EPA Region and request in the transmittal letter that the submission be parallel processed in order to expedite its incorporation into the SIP.

- **State proposes trading rule and solicits public comment.** The air agency would also move forward with publishing its notice of proposed rulemaking, consistent with the state's requirements. Once published, a public comment period would begin. A typical public comment period is 30 days; however, it may be possible to reduce the comment period to as little as 15 days under some circumstances.[31] We believe that a state air agency could complete these steps in about one month.

- **EPA reviews proposed trading rule.** The EPA Region would begin its formal review of the trading rule as soon as it receives the state's SIP submission. EPA's review and the subsequent publication of a notice of proposed rulemaking in the Federal Register would occur while the state was accepting public comments on its notice of proposed rulemaking and addressing any comments received. Because the EPA Region would have reviewed the draft through the interagency consultation process, the EPA Region should be able to complete its review of the submission quickly and prepare its notice of proposed rulemaking. We believe that an EPA Region could complete these steps in as little as one month if the state's rule closely follows the model rule and if the air agency thoroughly addressed comments offered by the Region during the interagency consultation process. If, however, the submission does not closely follow the model rule or if the Region's comments were not thoroughly addressed, these steps could take three months or more.

- **EPA proposes approval of the trading rule.** The EPA Region would then publish the notice of proposed rulemaking in the Federal Register. The publication of the notice would start a public comment period. A typical public comment period is 30 days; however, it may be possible to reduce the comment period to as little as 15 days under some circumstances. The EPA Region could complete these steps in about one month.

---

[31] Requirements for public hearings on SIPs can be found at 40 CFR 51.102. Section 51.102(g) allows a state to shorten its public comment period if it submits a written application to EPA in advance.

- **EPA addresses comments and prepares final approval.** At the close of the comment period on EPA's notice of proposed rulemaking, the EPA Region would address any comments received and could draft the Federal Register notice finalizing the approval of the trading rule; however, EPA cannot publish this final rulemaking notice until the state completes its rulemaking process and submits the adopted rule to EPA. We believe that the EPA Region could respond to any comments and draft the final rulemaking notice in about one month.

- **State prepares final rulemaking package.** By this time the state's public comment period would have closed. The state air agency would address any comments received and prepare its final rulemaking package, making any changes in the rule that are necessary to address any comments that are received. These actions would take about one month. It must be noted that if the state needs to make a significant revision to its trading rule, the EPA Region would not be able to complete its rulemaking through parallel processing because the state's final rule would not be consistent with the rule that EPA had proposed for approval.

- **State review, if required.** At this point some states may be required to submit the final rule for review within the state before submission to EPA for final approval. These reviews may be carried out by the state legislature, the Governor's office or other body established by the state. The length of these reviews varies greatly. In most cases these reviews either are not required or last less than six months. However, in some states the review can add up to 18 months to the process, particularly in cases where review is required by the state legislature and the rule is completed just after the end of one legislative session.

- **State submits adopted rule for EPA approval.** After any required state-level review is completed, the air agency would address any comments that resulted from this review process. The air agency would also complete the state rulemaking process which may involve steps such as publishing a final rulemaking notice. The state would then submit the final SIP package to the EPA Region for final approval. We believe that these tasks can be completed in about one month.

- **EPA finalizes approval.** The EPA Region would then complete its final rulemaking notice and have the notice published in the Federal Register. This step could take one month or perhaps less. EPA's final action would be effective 30 days after publication. If EPA approves the SIP and once EPA's rulemaking action is effective, the MPO(s) may use the trading rule in conformity determinations. No further rulemaking by the state or EPA would be necessary to allow trades to occur.

B.4 Has any state adopted a trading mechanism into its SIP this quickly?

Yes. The following timeline from Utah's experience illustrates how quickly a trading mechanism can be adopted into a SIP. In this case, the trade was between pollutants rather than between emissions sectors. The fact that the trading mechanism was critical for the area to demonstrate conformity meant that all of the involved agencies worked quickly to reach resolution:

- In early January 2002, the MPO for Salt Lake City, Utah, recognized that it would not be able to pass the NOx budget in its PM_{10} SIP in its next conformity determination that was required in July 2002. The MPO knew that it would be able to easily pass the PM_{10} budget in the SIP. Through interagency consultation, it was decided that the state and EPA would work together to quickly develop and approve a SIP trading mechanism that would allow the area to trade emission reductions from its PM_{10} budget to its NOx budget as allowed by 40 CFR 93.124(b).

- In early January, Utah and EPA began drafting the trading rule and supporting information. By early February, the initial language for the trading rule and supporting information had been completed. In early March, the regulatory language and supporting information was finalized.

- On March 15, the Governor of Utah submitted the proposed trading rule to EPA Region 8 and asked that parallel processing be used to approve the rule.

- On April 1, the Utah Air Quality Board (UAQB) proposed the trading rule and started a 30-day public comment period.

- On May 1, EPA Region 8 published a <u>Federal Register</u> notice proposing approval of the trading rule that started a 30-day public comment period.

- On May 13, the UAQB adopted the trading rule and submitted the adopted rule to EPA Region 8 for final approval.

- On July 1, EPA Region 8 published a <u>Federal Register</u> notice approving the trading rule into Utah's SIP.

- On July 31, EPA's approval became effective and the MPO was then able to use the trading rule in its pending conformity determination.

The entire process from the time that it was recognized that a trading rule was needed to the time that EPA's approval of the rule became effective was just under seven months. While the specifics of the trading rule developed for Salt Lake City are different than a trading mechanism for nonroad retrofit projects, EPA believes that it would be possible to develop and approve such a trading mechanism in a similar amount of time.

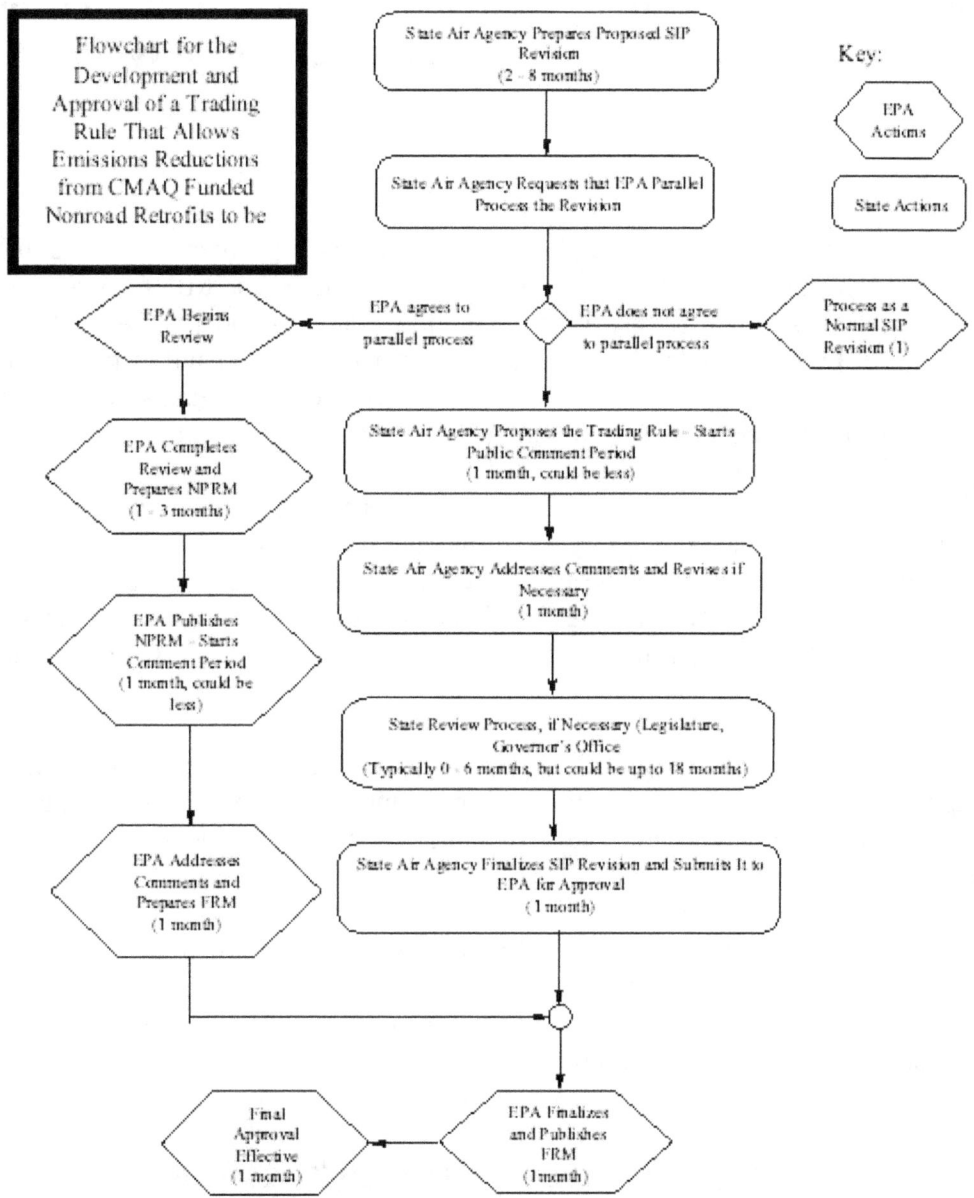

(1) EPA would not propose approval until after the state submits final SIP revision for approval. Would add 3 to 5 months to the process.

(2) Assumes any changes are minor and would not cause EPA to re-propose approval.

APPENDIX C:

Model Rule for Trading Emission Reductions from Nonroad Retrofit Projects for Transportation Conformity

Introduction

In recognition of the importance of nonroad diesel retrofit projects and other mobile source emission reduction strategies, MAP-21 continues to direct states and MPOs to give priority to funding diesel retrofits and other cost-effective mobile strategies under CMAQ. The law also notes that states and MPOs continue to have final CMAQ project selection authority. Furthermore, any state that has a $PM_{2.5}$ nonattainment or maintenance is required to invest a portion of its CMAQ funding on projects that reduce $PM_{2.5}$ emissions and precursor emissions.

This model rule establishes a trading mechanism that would allow emission reductions from nonroad retrofit projects that are eligible to be funded with CMAQ (23 U.S.C. 149(b)) to be used in transportation conformity. The model rule is intended to facilitate states' adoption of this type of trading mechanism into their SIPs. States would need to modify this model rule to account for any new issues raised by including projects that may not be eligible for CMAQ dollars, as described further below.

Once a trading mechanism is adopted by a state into its SIP, the state air agency, MPO, and other agencies as appropriate would follow its provisions each time the MPO wants to include the emission reductions from a new nonroad retrofit project (or set of projects) in a transportation conformity determination.

Appendix C is divided into two parts that follow a parallel structure. Part 1 provides background information for state and local agencies to consider for each section of the model trading rule. Part 2 provides the model rule itself that states can include in their SIPs. EPA developed the model rule based on existing laws and regulations.

Part 1: Explanatory Notes for Using Model Trading Rule Language

Section 1. Purpose

The purpose of the model rule is to establish a trading mechanism that allows certain nonroad retrofit emission reductions to be used in transportation plan and TIP conformity determinations.

Section 2. Definitions

Examples of definitions that may be needed are listed, but all of them do not need to be included if they are defined elsewhere in the state's regulations or if they do not apply (e.g., PM_{10} does not need to be defined if a state has no PM_{10} nonattainment or maintenance areas). At a minimum, the definitions of "CARB," "CMAQ," "EPA," "nonroad retrofit project," and "surplus" must be included. If you include a definition for a word or phrase that is defined in the conformity rule (40 CFR 93.101), your definition must be consistent with the definition found in that rule.

Section 3. Applicability

(a) This model rule could be tailored to establish individual trading mechanisms for more than one nonattainment or maintenance area within a state. Section 3(a) is intended to ensure that the exchange of emission reductions occurs within the geographic area that the regional emissions analysis covers, rather than between areas covered by separate regional emissions analyses. We believe that this language will cover all possible jurisdictional cases, e.g., where there is one MPO in a nonattainment area, an MPO that has a donut area, multiple MPOs in an area, or a multi-state area.[32]

States can apply this model rule to one or more MPOs, simply by including the names of the MPOs within the model rule. However, states could also list the MPOs to which the model rule applies and refer to this list in subsequent sections of the model rule. A state may want to choose this approach to be consistent with other parts of its SIP, or simply because doing so avoids repetition of the list in multiple places throughout the trading rule.

(b) As noted above, the model trading rule is intended to target retrofit projects that are eligible for CMAQ dollars. States would need to modify this model rule to account for any new issues raised by including projects that may not be eligible for CMAQ dollars. For example, a trading mechanism could also account for other retrofit projects that are funded or operated by a state or local transportation agency that may not be eligible for CMAQ dollars (e.g., retrofitting diesel mowing tractors that cut vegetation within the rights-of-way of roadway facilities).

(c) This section allows agencies to specify which pollutants and/or precursors are covered by their trading rule (e.g. the rule for an ozone area could specify that it applies to NOx, VOC, or both).

This section also states that the rule applies only to trades of the same pollutant or precursor. That is, the model trading rule does not provide for trading between pollutants or precursors.[33,34]

---

[32] Please refer to EPA's guidance, Guidance for Transportation Conformity Implementation in Multi-jurisdictional Nonattainment and Maintenance Areas (EPA-420-B-12-046), for more information regarding how the conformity rule's requirements for conformity determinations apply in areas that contain more than one MPO, a donut area, parts of more than one state, or any combination. See www.epa.gov/otaq/stateresources/transconf/regs/420b12046.pdf .

[33] If your SIP allows trading between pollutants and/or precursors that will be included in the retrofit trading rule, you will need to delete the sentence in Section 3(c) that says, "This rule applies only to trades of the same pollutant or precursor" and add a sentence similar to the following: "This rule does not interfere with [*name of state regulation*] which provides for trading between [*insert pollutants/precursors*]."

[34] If you are interested in allowing inter-pollutant or precursor trading for conformity purposes, you will need to do so within the SIP. Note that you would need to develop an appropriate trading ratio based on a technical justification and consider other factors as appropriate. Please consult your EPA Region if you intend to add inter-pollutant/precursor trading to your SIP.

Section 4. Applying Emission Reductions from Nonroad Retrofit Projects in Transportation Conformity Determinations

(a)(i) The interagency consultation process in the transportation conformity (40 CFR 93.105) would be followed for each nonroad retrofit trade. This part of the model trading rule would require the MPO, state department of transportation, other state and local transportation agencies (when appropriate), and the state air agency to consult on trades. Individual states can further clarify which agencies should be involved in their consultation process, and make any further modifications as appropriate.

(a)(ii) This section specifies that the MPO and the state air quality agency must agree on the quantity of emission reductions from a nonroad retrofit project, or set of projects, that can be used in conformity. The MPO and state air quality agency must work together for a trade to occur because each agency is responsible for part of the process. MAP-21 directs states and MPOs to give priority to projects that reduce $PM_{2.5}$ emissions, including diesel retrofits, in areas designated nonattainment or maintenance for $PM_{2.5}$ (23 US.C. 149(g)(3)). These projects may include both nonroad and on-road diesel equipment that are operated on highway construction projects within $PM_{2.5}$ nonattainment and maintenance areas (23 U.S.C. 149(k)(2)). Ultimately, the state DOTs and MPOs decide how to spend the CMAQ money they receive, in accordance with the CMAQ provisions of MAP-21. On the other hand, the state air agency is responsible for determining which emission control programs an area needs to meet its Clean Air Act obligations. The MPO and the state air quality agency must decide together how much of the emission reductions from a nonroad retrofit project or set of projects are appropriate to use in conformity. These two agencies would make a decision for each project or set of projects being considered for transportation plan and TIP conformity determinations.

State air quality agencies and MPOs need to consider the area's air quality needs when deciding how much of the emission reductions from a nonroad retrofit project is appropriate to use in conformity. For example, in the time period before a nonattainment area has an attainment demonstration, it may be appropriate to reserve some of the emission reductions generated by a nonroad retrofit project to help demonstrate attainment. Or, a state may decide to reserve some of the emission reductions for improving air quality, regardless of whether or not the attainment demonstration is in place. The state air agency and MPO could also decide that all of the emission reductions from a nonroad retrofit project can be used in the conformity determination. Whatever the case, the state air agency would document its concurrence in a letter to the MPO, which would describe the surplus reductions that are available for transportation conformity.

(b)(i) In order for emission reductions from a nonroad retrofit project to be estimated, NONROAD/NMIM would need the number of vehicles/engines/equipment being retrofitted, the vehicle/engine/equipment type and class being retrofitted, vehicle/engine/equipment model years, the retrofit technology being applied, the activity level of the vehicles/engines/equipment that are used (e.g., hours of usage), and when the retrofits will be implemented. This section requires an MPO to describe the details of the project. While an

MPO may not know all of this information with complete certainty, it should know enough about the project to make reasonable assumptions.

(b)(ii) The model rule relies upon the transportation conformity rule's requirements for ensuring that emission reductions from retrofit projects have sufficient commitments before they are accounted for in a regional emissions analysis. For example, retrofit projects that result from a state or local regulation or ordinance could be included in a conformity analysis once such a regulation or ordinance is adopted (40 CFR 93.122(a)(3)(i)). Conversely, a retrofit project that does not require a regulatory action to be implemented would meet this requirement if it is included in the transportation plan and TIP with sufficient funding and other resources for its full implementation. See Section 4.3 of this guidance for further information on what level of commitment is necessary to include retrofit reductions in a transportation conformity determination.

Whatever the case, once an entity (e.g., an owner or operator of nonroad equipment or vehicles) provides a commitment for implementing a nonroad retrofit project and the MPO relies on it in a conformity determination, it is an enforceable obligation. That is, under the existing provisions of the transportation conformity rule, an entity that makes a written commitment is subject to civil action if the entity does not fulfill its commitment.

(b)(iii) - (b)(v) These paragraphs are similar to the conformity rule's requirements for any control measures that are relied upon in a conformity determination.

(c) Emission reductions must be quantified using the latest assumptions available at the start of the regional emissions analysis (40 CFR 93.110). The consultation process would be used to evaluate and choose the model(s) and associated methods and assumptions to be used (40 CFR 93.105(c)(1)(i)). If the emission reductions are calculated using a method that relies on inputs such as temperature that were also used for developing the budgets in a SIP, these factors must be consistent with those used to establish the SIP as required in 40 CFR 93.122(a)(6). Quantification methods and information on the efficacy of retrofit technologies are addressed in more detail in Section 2 of this guidance document.

(d) This requirement for documentation is similar to the conformity rule's requirements for documenting the transportation plan, the TIP, and any control measures that are included in the regional emissions analysis of a conformity determination. The MPO would document in its conformity determination that the state air agency has concurred on the traded nonroad retrofit emission reductions. This documentation could be completed by referencing the state air agency's concurrence letter in the transportation plan/TIP conformity determination.

(e) In subsequent determinations, the MPO and state air agency would follow the consultation procedures in 40 CFR 93.105, but would not have to renegotiate the amount of available emission reductions from a nonroad retrofit project that has already been traded for a given year of the regional transportation conformity analysis. However, the MPO must recalculate the emission reductions that result from the project in subsequent conformity determinations. The MPO would also cite and/or include the previous air agency concurrence letter in its new conformity determination.

For example, suppose an area that is nonattainment for the 2008 ozone NAAQS has a trading mechanism in place, and the MPO is determining transportation plan/TIP conformity in the year 2015. The MPO determines that a new nonroad retrofit project reduces NOx by 5 tons per day in 2015 and 2 tons per day in 2020. After discussion in the interagency consultation process, the MPO and state air agency agree that these reductions are surplus, and the MPO can use some of the reductions in the years 2015 and 2020 of the regional emissions analysis for the transportation plan/TIP conformity determination. Specifically, in this example, the MPO and state air agency agree that the following reductions can be used for conformity purposes: 3 of the 5 tons per day of NOx reductions for the 2015 conformity analysis year and all 2 tons per day of NOx reductions for the 2020 conformity analysis year.

The next time the MPO does transportation plan/TIP conformity, the MPO re-calculates the retrofit project's reductions based on the latest models and assumptions, and it is found that the project reduces emissions by 6 tons per day in 2015, and 3 tons per day in 2020. In this example, the MPO can continue to use up to 3 tons per day of reductions in 2015 and up to 2 tons of reductions in 2020, based on the previous trade for this nonroad retrofit project. If additional reductions are needed from this project for any years, then the trading mechanism would need to be used to conduct a new trade.

As with any control measure that an MPO includes in its conformity determination, the MPO can only use the emission reductions from the project, or part of the project that is actually occurring. If the nonroad project's implementation is delayed, the MPO cannot include emission reductions from the project until its implementation is assured.

Section 5. Prohibition on Double-Counting

This section is intended to ensure that there is no double-counting of emission reductions that have already been used in a transportation conformity determination or in meeting any other Clean Air Act purpose. The best way to ensure this is to require that the impact of any trades be accounted for in any subsequent inventory analyses that are done. For example, if the application of this rule results in a decrease in the nonroad emissions inventory and an increase in the allowable on-road emissions, future SIP inventories or regional transportation conformity analyses should reflect those changes as appropriate.

Part 2: Model Trading Rule

Section 1. Purpose

This rule establishes the procedures that may be used to trade emission reductions from nonroad retrofit projects, as defined below, to the transportation sector for the purpose of determining conformity of a transportation plan or transportation improvement program.

Section 2. Definitions

The following definitions apply to this rule:

- *CARB* means the California Air Resources Board.

- *CMAQ* means the Congestion Mitigation and Air Quality Improvement Program, as defined in title 23, U.S.C.

- *EPA* means the U.S. Environmental Protection Agency.

- *Nonroad retrofit project*, for the purpose of this rule, means an undertaking to reduce emissions from nonroad vehicles or engines as described by 23 U.S.C. 149(b), below the emissions level which is currently required by EPA regulations at the time of vehicle or engine certification. For the purposes of this rule, such an undertaking must apply a technology verified by EPA, CARB, or other entity recognized by EPA for verifying retrofit technology, use an EPA-certified engine replacement or early replacement of older vehicles or equipment with cleaner vehicles or equipment; and it must be eligible for funding under CMAQ.

- *NOx* means oxides of nitrogen.

- PM_{10} means particulate matter that is less than or equal to 10 microns in aerodynamic diameter.

- $PM_{2.5}$ means particulate matter that is less than or equal to 2.5 microns in aerodynamic diameter.

- *Surplus* means that emission reductions are not otherwise relied on to meet any Clean Air Act air quality related purpose including but not limited to reasonable further progress, attainment, maintenance, or requirements adopted to satisfy Clean Air Act section 110(a)(2)(D). In the event that a nonroad retrofit project is relied on to meet such an air quality requirement, emission reductions are no longer surplus and may not be used in transportation conformity determinations.

- *VOC* means volatile organic compounds.

Section 3. Applicability

(a) Geographic applicability: This trading rule applies to the geographic area covered by the regional emissions analysis done for a transportation conformity determination for a transportation plan or transportation improvement program in *[insert name of nonattainment or maintenance area]*.

(b) Project applicability: This trading rule applies to nonroad retrofit projects as defined in Section 2 of this rule.

(c) Pollutant applicability: This trading rule applies to the following pollutants/precursors: *[insert pollutants/precursors; for precursors indicate for which pollutant and standard(s)]*. This rule applies only to trades of the same pollutant or precursor.

Section 4. Applying Emission Reductions from Nonroad Retrofit Projects in Transportation Conformity Determinations

Before *[insert name of MPO]* can include emission reductions from a nonroad retrofit project(s) in a transportation conformity determination, the steps in paragraphs (a) – (d) must be completed.

(a) Interagency Consultation.

    (i) *[Insert name of MPO]*, *[insert name of state department of transportation]*, *[insert names of other state and local transportation agencies, when appropriate]* and *[insert name of state air agency]* shall follow consultation procedures in *[insert "40 CFR 93.105" or state transportation conformity SIP if one has been approved by EPA]* throughout the implementation of this rule.

    (ii) *[Insert name of MPO(s)]* and *[insert name of state air agency]* must concur on the amount of emission reductions from a nonroad retrofit project(s) that can be used in the transportation conformity determination. *[Insert name of state air agency]* will document this concurrence in a letter to *[insert name of MPO]*. Concurrence of *[insert name of state air agency]* also affirms that the emission reductions are surplus and therefore available for the transportation conformity determination.

(b) Description of the Nonroad Retrofit Project(s).

    (i) *[Insert name of MPO]* must fully describe each nonroad retrofit project being relied upon in a conformity determination, including the number of vehicles/engines/equipment being retrofitted, the vehicle/engine/equipment type and class, vehicle/engine/equipment model years, the retrofit technology being applied, the activity level of the vehicles/engines/equipment, and the implementation schedule of the nonroad retrofit project.

    (ii) *[Insert name of MPO]* must provide assurance that the nonroad retrofit project is enforceable by ensuring that any nonroad retrofit project under this rule meets the requirements of 40 CFR 93.122(a);

    (iii) *[Insert name of MPO]* must provide assurance that the nonroad retrofit project has adequate funding and resource commitments to ensure that emission reductions from the nonroad retrofit project will occur in the years of the regional emissions analysis when emission reductions will be used;

(iv) *[Insert name of MPO]* must provide assurance that the nonroad retrofit project is on schedule and that the retrofitted vehicles, engines, or equipment will remain in use within the nonattainment or maintenance area in the years of the regional emissions analysis when the emission reductions will be used; and

(v) *[Insert name of MPO]* must provide assurance that emission reductions will be based only on activity that occurs within the geographic area covered by the regional emissions analyses.

(c) Calculation of Emission Reductions. For each year of the regional emissions analysis in which emission reductions from the nonroad retrofit project(s) will be applied, *[insert name of MPO]* and *[insert name of state air agency]* must calculate emission reductions from the nonroad retrofit project(s) using:

(i) the latest EPA-approved emissions model or other method as determined through the interagency consultation process required by 40 CFR 93.105;

(ii) data and assumptions consistent with requirements for use of latest planning assumptions under 40 CFR 93.110, including, but not limited to current data or future projections of numbers and types of nonroad vehicles/engines/equipment being retrofitted, and current data or future projections of hours of use for those vehicles/engines/equipment within the nonattainment or maintenance area;

(iii) the latest verified information available regarding the efficacy of the nonroad retrofit project as provided by EPA and/or CARB; and

(iv) ambient temperatures and other relevant factors consistent with those used to establish the motor vehicle emissions budgets (if they exist) in the applicable implementation plan, as required by 40 CFR 93.122(a)(6), if a method is used that requires such factors as inputs.

Once total emission reductions from the nonroad retrofit project(s) are calculated, *[insert name of MPO]* can use no more than the amount of emission reductions determined in Section 4(a)(ii) to be available for the conformity determination.

(d) Documentation. *[Insert name of MPO]* must document in the transportation conformity determination how steps (a) through (c) of this section have been satisfied, consistent with the transportation conformity regulations at 40 CFR Parts 51 and 93.

(e) Subsequent conformity determinations. Once emission reductions from a nonroad retrofit project(s) have been used in a conformity determination, *[insert name of MPO]* can include emission reductions from those projects for the same analysis years in a subsequent conformity determination without repeating paragraph (a)(ii), provided *[insert name of MPO]* meets all other requirements of Section 4 of this rule.

Section 5. Prohibition on Double-Counting

Once the emission reductions of the nonroad retrofit project(s) are included in a transportation conformity determination, those specific reductions are no longer surplus and therefore no longer available as new reductions for meeting any Clean Air Act air quality related purpose, including but not limited to, reasonable further progress, attainment, maintenance, or requirements in Clean Air Act section 110(a)(2)(D). Any emissions inventory created after a trade must properly account for the emission impact of the trade.

www.ingramcontent.com/pod-product-compliance
Lightning Source LLC
Chambersburg PA
CBHW081737170526
45167CB00009B/3850